THE NATURAL HISTORY MUSEUM BOOK OF

GEMSTONES

A CONCISE REFERENCE GUIDE

ROBIN HANSEN

Published by the Natural History Museum, London

To Chris, Kaylee and Devon for your love, help and patience during the many hours spent writing this book, I hope 'these facts will blow your mind', and to Leonie and Richard for the constant love, encouragement and advice, and for giving me my love of gems.

First published by the Natural History Museum,
Cromwell Road, London SW7 5BD

© The Trustees of the Natural History Museum, London, 2022
Reprinted with updates 2023

ISBN 978 0 565 09224 5

A catalogue record for this book is available from the British Library

10 9 8 7 6 5 4 3 2

Designed by Mercer Design, London
Reproduction by Saxon Digital Services
Printed by Toppan Leefung Printing Limited, China

Front cover: top row, left to right: tourmaline, tourmaline, padparadscha sapphire, kyanite, sphalerite; middle row, left to right: tourmaline, amethyst, synthetic spinel, sapphire, black opal; bottom row, left to right: synthetic emerald, malachite, peridot, tourmaline, citrine.
Back cover: sapphire set in a button of rock crystal inlaid with gold, rubies and emeralds. From the Sir Hans Sloane Collection, the Natural History Museum, London.

Contents

1 Introduction

Gems have been objects of desire for millenia, with historical, religious, spiritual and scientific significance. They are symbolic, representing power, superstition, loyalty and romance. Gems have been used as amulets and talismans to give protection or bring power, believed to have magical and healing powers. They are used to adorn and decorate, enjoyed for their splendour or given purpose as carved vessels and seals. This book takes you on a journey to discover the history of gems, how they each form, their distinguishing features and properties, and how they are cut and used. I hope to inspire a greater appreciation of their beauty and diversity, and ignite an appetite for further learning. Gemmology has something for every interest, crosscutting many disciplines from chemistry and physics to history and the arts.

What is a gem?

There is no single accepted definition of a gem, but without doubt it is something that is treasured, usually fashioned to reveal its beauty, and used for adornment. The definition can be expanded to include worked objects, that is gem materials fashioned into ornaments which are not worn, such as carvings, spheres and bowls. There is no limit to what can be used – minerals, rocks, fossils, organic or human-made materials – as long as they are cherished.

▼ The colour and brilliance of this faceted 30 ct yellow sapphire accentuate its beauty.

◄ Sapphire carved as a Buddha, mounted on a gold pin.

◄ The De Beers Millennium Star Diamond (203.04 ct) is perfectly cut for beauty, taking over three years to fashion.

Gemstones, normally considered as those used in jewellery, have three key attributes: beauty, rarity and durability. Each gem has a unique beauty defined by the way it interacts with light – a combination of colour, transparency, lustre and sparkle. Almost all are polished, or cut and faceted, to reveal this beauty to its full effect. Rarity makes gems desirable, and people aspire to possess or wear them. Durability means they are robust enough to be worn, with their beauty surviving the test of time, although for some, such as pearl, the beauty outweighs their fragility. Gemstones are commonly divided into two categories: diamond (both colourless and fancy colour) and coloured stones (everything except diamond, even if colourless). This is because the two have very distinct industries, from their mining and supply, to cutting methods, style of fashioning, grading and selling.

There are numerous factors beyond beauty, rarity and durability that determine a gem's desirability. They are valued financially, socially, culturally, emotionally and scientifically. Their size, optical phenomena, geological and geographical source, folklore and popularity all influence their worth. These treasures were once reserved for royalty or the rich, especially those in limited supply, and passed as heirlooms from generation to generation. Today gems are widely traded and enjoyed. The development of treatments, imitations and human-made synthetics means gems are now available to fit all budgets, even the most coveted diamond, ruby, sapphire and emerald.

▲ Gemstones are fashioned from many different materials, most commonly minerals.

Most natural gems are formed by geological processes deep beneath Earth's surface, over millions of years. The vast majority are minerals, defined as natural, inorganic solids with a set composition of chemical elements arranged in a regularly repeating atomic structure. It is the unique combination of composition and crystal structure that gives each gem its distinctive properties. Some gems are rocks, which are aggregates of one or more minerals, intergrown to create colourful textures or high durability. Rocks commonly used as gems include lapis lazuli and jade. Organic gems have biological origins, that is they are made from, or by, plants or animals. Some, such as pearl and coral, consist partly of minerals and are more correctly termed biogenic. Others contain only organic material, such as amber and jet.

The scientific value of a gem lies with the information it holds. Gems that form over long periods of time can tell us about the history of Earth. Diamonds originate hundreds of kilometres beneath the surface, providing a snapshot into our inner planet, and Earth's early formation. Amber is similarly a time capsule, but for Earth's surface, a fossilized tree resin that preserves plant and animal life from millions of years ago. The discovery of many scientific phenomena came from the analysis of gem minerals. Fluorescence,

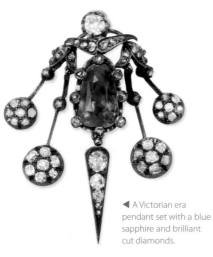

◀ A Victorian era pendant set with a blue sapphire and brilliant cut diamonds.

a type of luminescence, was observed in fluorite, from which it is named. Tourmaline, used in the early 1700s for its ability to attract warm ashes from pipes, was first noted for the properties of pyroelectricity and piezoelectricity. Rare blue diamonds are the only diamonds to conduct electricity, and the discovery that boron impurities caused both the colour and the conductivity led to the development of boron-doped synthetic diamond, used as a superconductor with outstanding properties. Synthetic gem materials are grown as large pure crystals, and the majority are in fact used for industry.

► Diamonds have only reached their height in popularity in the last few centuries, following the ability to facet them into brilliant gemstones.

The age of a gemstone is perhaps one of the most fascinating, but overlooked, facts. Diamond is prized for its brilliance and fire, but few people realize the sparkling jewel they are wearing is at least 660 million years old, if not several billion. Ruby, sapphire, emerald and amber are likewise millions of years old. At the other end of the spectrum, as humans have learned to harness nature, pearls can be cultivated in just six months, and amazingly, a ruby can be grown synthetically in less than a day.

◄ Turquoise has been used as a gem for millennia, for its colour and easiness to fashion.

As well as telling us about the history of Earth, the use of gems through the ages illuminates human history. Gems provide insight into past societies and their culture, defining what they valued. The geographical distribution of gems from their source also maps out trade routes over time.

Possibly the oldest known gems are beads made from shells, with evidence of use over 100,000 years ago. Jade has been worked for its unbreakable nature for at least 7,000 years, highly treasured in China and by the Olmecs in Mesoamerica. Gems such as peridot, lapis lazuli, carnelian and turquoise are found in the jewellery of Ancient Egyptians, Greeks and Romans, with garnet prized as far back as 5000 BC. Emerald, amethyst, agate, as well as amber, jet, pearl and coral have similarly long histories. These gemstones were not only viewed as objects of beauty but were symbols of status and wealth, amulets to protect, and were believed to have magical powers to cure.

The history of gem names, use and source is documented in literature. The first written record of gemstones in the western world is credited to the Greek philosopher and naturalist Theophrastus (Hellenistic era c.371–287 BC) with his treatise *Peri Lithon* (On Stones c.314/315 BC). He was the first to mention the phenomenon of pyroelectricity, possessing an electrical charge upon heating, noting that a gem (thought to be tourmaline) attracted pieces of straw when warmed. Roman historian, naturalist and philosopher Pliny the Elder drew on the work of Theophrastus for his own publication *Historia Naturalis* (Natural History, completed 77 AD), describing around 300 stones. These texts were used as standard works for over 1,000 years, and Pliny is the most quoted historical author on gems and minerals. Ancient East Asia and the Old Testament have their own records. The Book of Exodus describes the Breastplate worn by the High Priest of the Temple, which contained 12 sacred stones. These represented the 12 tribes of Israel, later associated in the Middle Ages with the 12 zodiac signs, and in modern times with the birthstones of the 12 months of the year.

▲ A modern version of the 12 birthstones (from left to right and top to bottom): January, garnet; February, amethyst; March, aquamarine; April, diamond; May, emerald; June, pearl; July, ruby; August, peridot; September, sapphire; October, tourmaline; November, citrine; December, tanzanite.

The interpretation of past literature is not without challenge. Original manuscripts may no longer exist. There are also multiple translations and interpretations, in which meanings are unwittingly altered and reiterated, allowing errors to become fact. The etymology of many gem names is long and complex – names changing over time and with modern gems retrospectively fitted to ancient descriptions there are multiple gems for one name. Topazion, for instance, translated as topaz, was the name applied to any golden gem including peridot, however modern topaz may not have been known until the 1700s.

The Middle Ages (~500 to 1300 AD) focused on the healing and magical properties of gems. It was not until the Renaissance that their descriptions became more scientific. Gemstones were distinguished by their appearance and physical properties including colour, transparency and hardness. Similarly coloured gemstones were assumed to be the same, such as ruby and red spinel.

Diamonds, traded from India, found their way into European jewellery from the 1300s, and with this came the development of cutting centres. The first recorded use of a diamond in an engagement ring was in 1477. European explorers, setting out to discover new trade routes, returned in

the 1500s with jade and emerald from the New World, while locally garnet from Bohemia and agate from Idar-Oberstein rose in popularity. In the 1700s tourmaline arrived from Sri Lanka, and the discovery of diamonds in Brazil replaced the dwindling sole source, India. Discoveries in the 1800s of amethyst and agate deposits in Brazil and Uruguay, opal in Australia, and diamond in South Africa, expanded the gem trade globally.

The 1700–1800s were a turning point in the knowledge of gems. Mineralogy emerged as a science, the chemical elements were discovered, and study of crystals and their structure meant minerals could be identified and classified, with their properties defined. Ruby and sapphire were brought together, recognized as one mineral corundum, while Balas ruby was given its own identity as spinel. Emerald was found to be a variety of beryl, and jasper, carnelian and agate, so long having their own identities, became quartz.

Since the turn of the twentieth century, knowledge and technology have advanced at a dramatic rate, bringing with them the ability to commercially synthesize gems. New gems, and gem varieties, have been discovered, including kunzite, morganite and tanzanite. Improved analytical techniques have further separated minerals, such as the rare gem sinhalite from peridot. Today, analysis at atomic levels provides confident identification of deceptive treatments and synthetics, and can indicate a gem's origin through signatures of impurities.

With identification now so precise, there are international standards in nomenclature, despite the multitudes of trade and marketing names. The International Mineralogical Association (IMA) manages the approved list of mineral species and their names. Other bodies such as CIBJO, The World Jewellery Confederation, regulate gem definitions and standards in the trade to prevent misleading names, and ensure that treatments, imitations and synthetic gemstones are appropriately disclosed.

How and where do gems form?

Most gem materials, such as diamond, ruby and emerald, are minerals. Others, including amber and pearl, have biological origins. For a mineral to crystallize, it requires several things: the right ingredients (the essential combination of chemical elements), the necessary geological conditions (e.g. pressure and temperature), and enough time, and room, to grow.

To understand how and where gem minerals form, there must be an understanding of Earth, its structure, and where different elements are found. When Earth formed around 4.5 billion years ago, the heavier elements sunk slowly to the centre and the lighter elements drifted up to the surface. They eventually separated into distinct layers – a central dense core thought to be iron and nickel, around that a mantle of magnesium and iron silicate minerals, and finally a thin crust on the top, rich in silicon, oxygen, potassium, sodium and aluminium.

The majority of gem minerals form within Earth's crust. The crust is predominantly made of silicate minerals, with more than half composed of the feldspar group and quartz. The growth of gem

minerals, especially non-silicates, is constrained to where their necessary elements are found. Other gem minerals, such as diamond, garnet, chrome diopside and peridot, form below the crust in the upper mantle, and are brought to the surface through volcanic activity. On very rare occasions gem minerals such as peridot come from outer space in pallasite meteorites.

Earth has two types of crust: the thicker (25–70 km, 16–43 miles) continental crust rich in silicon, aluminium and oxygen, and the thinner (6–12 km, 4–7 miles), denser oceanic crust higher in magnesium and silicon. Beneath the crust is the uppermost mantle, a rigid portion which, together with the crust, forms the lithosphere. The lithosphere is fractured into plates that 'float' on a more ductile mantle layer beneath called the asthenosphere. These plates gradually move over time, in a process known as plate tectonics, believed to have occurred for the past three billion years. During this time the continents have changed dramatically, drifting into different formations. The movement of the plates, particularly the formation and break-up of supercontinents such as Gondwana and Pangaea, is responsible for the formation of many gem minerals found today.

These large-scale tectonic events allow different elements to come together, and create the geological conditions required for the formation of new minerals. When two plates of continental crust collide, they squeeze together pushing up huge mountain ranges. This action generates intense heat and pressure with partial melting, leading to the upwelling of molten rock (called magma), volcanic activity and metamorphism. This creates the conditions in which gem minerals such as ruby, sapphire and spinel form. When

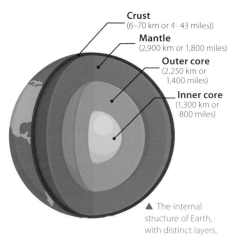

Crust
(6–70 km or 4–43 miles))
Mantle
(2,900 km or 1,800 miles)
Outer core
(2,250 km or 1,400 miles)
Inner core
(1,300 km or 800 miles)

▲ The internal structure of Earth, with distinct layers.

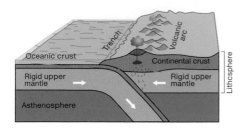

▲ The collision of two tectonic plates causes subduction of the oceanic crust under the continental crust, resulting in a chain of volcanoes.

plates of continental and oceanic crust collide, the denser oceanic crust dives beneath the lighter continental crust, a process called subduction. As it is forced down it loses water, causing the mantle to melt. The magma rises to the surface, forming chains of volcanoes. Jade forms in subduction zone settings. When two plates move apart, rifting occurs and new crust is formed. Associated volcanism brings gem minerals such as ruby and sapphire to the surface.

The Pan-African Orogeny (750–450 million years ago) was a series of events related to the assembly of Gondwana, with a complex history of tectonic activity including deformation, metamorphism and magmatism. This brought East Africa together with Madagascar, Sri Lanka and southern India. It created the Mozambique Orogenic Zone, a huge geological structure running north–south from Ethiopia and Sudan to Mozambique and Madagascar. This is one of the richest areas in the world for gem deposits, known as the East African gemstone belt. Gem minerals include tanzanite, garnet, spinel, emerald, ruby and sapphire. Madagascar and Sri Lanka have related geological histories with countless episodes of mineralization, and both are famous for their incredible wealth of gem deposits. The assembly of Gondwana also brought together the east of South America, and the west of Africa. The collision of these plates and associated magmatism is responsible for the

formation of the extensive pegmatite province in northeast Brazil, renowned for its gemstone production including tourmaline, beryl and topaz.

The collision of the Indian and Eurasian tectonic plates (15–5 million years ago) caused the formation of the Himalayan and Alpine mountain ranges. As well as creating high pressure and temperature conditions favourable for gem formation, it also pushed deposits towards the surface. This created the gem-hosting marble deposits through Afghanistan, Pakistan and Myanmar to Vietnam, as well as the Kashmir sapphires.

Geologists classify rocks into three types: igneous, metamorphic and sedimentary. Together these different types form a continuous evolution known as the rock cycle.

Igneous rocks form from the cooling and solidifying of magma. Intrusive igneous rocks form at depth, through the upwelling of magma in the crust, cooling slowly to allow the formation of larger crystals. Extrusive igneous rocks form on the surface, through volcanic action, cooling quickly into fine-grained rocks. Igneous rocks are further classified by their composition. **Felsic**, named from **fe**ldspar and **si**lica, are high in silica (SiO_2), with aluminium, sodium and potassium. They are generally light coloured and common in continental crust. **Mafic** rocks are lower in silica and are richer in **ma**gnesium, iron ('**ferrum**' in Latin) and calcium. They are darker in colour

◄ Fire opal filling cavities in rhyolite, a felsic volcanic (extrusive igneous) rock.

and denser than felsic rocks, and common in the oceanic crust. Ultramafic rocks are very low in silica, and are found deep underground in the upper mantle.

Intrusive felsic rocks include granite, which hosts feldspar, quartz, zircon and topaz. Intrusive mafic rocks include gabbro, and ultramafic rocks include peridotites and eclogites, which are associated with peridot, diamond, garnet and chrome diopside. Another intrusive igneous rock is nepheline syenite, which hosts sodalite. Extrusive igneous rocks are generally too fine-grained to form gem minerals, however they bring gems such as sapphire and diamond to the surface through volcanic activity. As the lava cools, trapped gas bubbles leave cavities in which gem minerals can crystallize. Felsic volcanic rocks, like rhyolite, are associated with topaz, red beryl and fire opal, while mafic volcanic rocks, such as basalt, host the quartz varieties amethyst, agate and jasper.

Pegmatites are a special type of igneous rock with large crystals, and are one of the main sources of gem minerals. They contain rare elements in high enough concentrations to allow certain gem minerals to grow including beryllium (beryl, chrysoberyl), boron (tourmaline), lithium (spodumene), and fluorine (topaz, fluorite). Pegmatites form in the final stages of crystallization of large magma intrusions. As the main body slowly cools it crystallizes into granite, containing the common rock-forming minerals of quartz, feldspar and mica. Watery fluids and rare elements that occur in low concentrations throughout the magma are not incorporated into these common minerals, and become progressively concentrated, rising with the magma. In the last stage of crystallization, the remaining melt becomes trapped in pockets in the upper parts of the granite, or seeps into cracks in the rock surrounding the intrusion. This melt crystallizes to form pegmatites, and it is thought

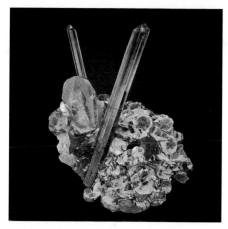

▲ Colour-zoned tourmaline crystals, 10.3 cm high, intergrown with lepidolite mica and quartz from a pegmatite in Barra do Salinas, Minas Gerais, Brazil.

its watery, enriched nature allows individual crystals to grow rapidly, often to extraordinarily large sizes. Magnificent transparent gem crystals form in cavities. As the cooling and crystallization continues within these pockets, using up the elements, the composition of the fluid changes. Crystals that have already formed may be etched, seen in beryl, topaz and spodumene, or have multiple generations of growth creating zoning as seen in tourmaline.

Metamorphic rocks have undergone heat and/or pressure, causing them to change and recrystallize without melting, forming new minerals and textures. Regional metamorphism occurs on a large scale through tectonic events such as the collision of continents. Contact metamorphism occurs locally from magma intruding and heating the surrounding rocks. When fluids are involved, chemically altering the surrounding rocks in situ, this is known as metasomatism.

Gems formed through regional metamorphic processes include corundum, garnet, iolite, emerald, kyanite, and jade. Gems formed by contact metamorphism with associated fluids include emerald, chrysoberyl and lapis lazuli.

▼ Ruby crystals in the metamorphic rock gneiss, from Mysore, India.

▶ Hessonite garnet in limestone skarn from Massachusetts, USA.

▼ Conglomerate, a sedimentary rock made of rounded fragments, hosting a diamond crystal, from Diamantina, Brazil.

Metamorphic rocks are classed as low, medium or high grade, depending on the pressure and temperature conditions reached. The type of metamorphic rock is also determined by the composition of the original rock. Limestone forms marble, granite forms gneiss, mafic rocks form amphibolite, and mudstones or shales can form schist. Serpentinites are rocks formed from the alteration of ultramafic rocks by low temperature metamorphism with associated fluids.

Hydrothermal (hot watery) fluids play an important role in the formation of gem minerals in both igneous and metamorphic settings. As hot fluids move through the crust, they dissolve minerals and leach elements in some areas, transporting them to other areas. New minerals crystallize into cracks and fractures as hydrothermal veins, including quartz, rhodochrosite and fluorite. The circulation of fluids can unite rare elements not normally found together, such as chromium and beryllium, allowing the formation of alexandrite and emerald. They also cause the formation of amethyst and agates in cavities in basalt. Fluids responsible for metasomatism form nephrite jade. Contact metamorphism combined with metasomatism produces a skarn at the boundary between limestone (a carbonate rock) and

intruding granite. The silica-rich fluids react with the carbonate rocks altering them to calc-silicate minerals, allowing gem minerals such as rhodonite, garnet and diopside to grow.

Sedimentary rocks form through the weathering and erosion of other rocks, breaking them down into fragments that are accumulated, compacted and cemented together to form a new rock. They are usually layered, with younger fragments deposited on top. The fragments may be angular or rounded, and can be of any size. Sedimentary rocks often consist of more durable minerals such as quartz. Sandstones are composed of sand-sized grains. Limestones are made of calcium carbonate minerals and can form through the accumulation of shell and coral pieces. Shale and mudstones are very fine-grained rocks.

▼ Washed gem gravel containing ruby and spinel, from Mogok, Myanmar.

eluvial (washed away by water), aeolian (blown and sorted by wind) and alluvial (transported by water through drainage systems such as rivers and streams). Alluvial deposits, such as the gem gravels of Sri Lanka and Myanmar, are the most important, producing more gems than any other type of primary or secondary deposit. More than a quarter of all gem minerals can be found in placer deposits, including ruby, sapphire, topaz, spinel, garnet, tourmaline, zircon, beryl, chrysoberyl and many rare gem minerals.

Fluids from groundwater, potentially mixed with hydrothermal fluids, can cause the formation of gem minerals near Earth's surface. Rainwater seeps into the ground and interacts with minerals, dissolving and transporting elements. When these reach areas of particular geological conditions, they crystallize out of solution forming new minerals. Silica-rich fluids form opal, and fluids bearing copper can form azurite, malachite and turquoise.

Primary deposits are those where the gem mineral is found in situ, and extracted from the rock in which it formed. While the crystals are well preserved in this type of deposit, the concentration is usually low and often sporadic, with a range of quality. These deposits tend to be in hard rock and are challenging to mine. They may require the removal of tonnes of material using heavy machinery and explosives, which risks damaging the crystals. These deposits also require an understanding of the local geology and formation processes to mine them effectively. Gemstones that have low durability such as emerald and tanzanite are mined from primary deposits.

Secondary deposits are where the gem minerals have been weathered and eroded from their primary source, then transported and deposited elsewhere. Known as placer deposits, these are classified by the transport and deposition process: colluvial (transported by gravity down a slope),

Secondary deposits are easier to mine as they are close to the surface and often unconsolidated. They can be mined by hand without heavy machinery or explosives. While the crystals may be rounded or waterworn, the grades are higher as they are naturally sorted along the way. Included and fractured stones get broken up while higher quality crystals remain. This process also sorts the larger from the smaller, the higher density from the lower, and the harder, tougher gems from the less durable. This natural sorting process can be used to locate primary deposits. Olivine is one of the gem minerals most affected by chemical weathering, so is only found in placer deposits close to the primary source. Diamond, famed for its incredible hardness and stability, can be transported so far that the primary source cannot be found. As olivine can occur with diamond, it is therefore used as an indicator mineral for a diamond primary deposit.

Some gems are formed by animals or plants. Biogenic gems result from the activity of living organisms. Pearls and shells are produced by molluscs. Coral is formed from the skeletons of tiny marine animals called polyps, in colonies of branching structures. Ivory is the tooth (including tusks) of several different mammals such as elephants. Amber and jet are organic gems produced from tree resin and wood, respectively, which have been fossilized by the sedimentary processes of heat and pressure through burial.

Sourcing gems

Every gemstone is unique, varying in appearance, quality and size, and has its own pathway from its source to the consumer, or in the case of synthetic gem materials, from a factory to the consumer.

This journey is called the gemstone pipeline. Many gems have a straightforward journey from mine to market. Once extracted in their natural unfashioned state – termed 'rough' – they are sorted, graded, sold to a dealer in gem rough, sent to a cutting centre, sold to a gem dealer or jewellery manufacturer, then a retailer and finally the consumer. The cost of a gemstone will generally increase with each step of the journey. Other steps in the pipeline may include treatment of the rough or finished gem, assessment by gem laboratories, valuations and appraisals. The journey continues with the consumer, with gemstones passed down through generations, or returned to market through antique dealers, pawnbrokers or auction houses.

The gemstone pipeline for diamond is well documented. Most mining is undertaken by a small number of international companies, some of whom control the entire journey including cutting and retailing. The mining is larger scale, more mechanised, and more secure than any other gem. There are strict controls on the import and export of rough, well-defined grading systems, and worldwide pricing guidelines. Coloured stones have a much more varied pipeline. Up to 80% of mining activity is undertaken by artisanal and small-scale miners, and the gem may pass through many more hands before reaching the consumer.

Certain places are perceived as producing the finest gemstones, which can dramatically add value, for example Kashmir sapphire, Burmese ruby or Colombian emerald. The desire to know where a gem comes from is also important for political, ethical, economic and environmental concerns. Consumers, particularly the millennial generation, are more aware of the processes to extract gems and wish to know their gemstone was mined responsibly and sustainably. Was it legally mined, obtained and exported following local as well as international laws and regulations? Did the sale of the gem fund terrorism, or involve money laundering? Did the mining impact on the environment? Did the miners have good working conditions, or was there child labour? Did the mining have a positive impact on the local economy providing employment? Some gemstones retain their history, or are recognized as being from one source, such as tanzanite. But for others, despite the importance of traceability, the numerous steps in their journey result in the loss of their provenance. Furthermore, there have been huge advances in treatment methods, and the production of synthetic gem materials. While this increases the affordability and availability of different gem materials, it can drive up prices of natural, untreated gems, and also change the public's perception of a gem. Inadequate disclosure of treatments and synthetics, whether intentionally or unknowingly, breaks the trust of the buyer and is damaging to the whole industry.

Consumers today expect transparency in the market, and demand proof of authenticity, especially for high-end gemstones. Reputable gem laboratories evaluate gemstones to confirm identity, signs of treatments, and give opinions on country of origin. Determining geographical origin is not simple as similar geological deposits produce similar gemstones, even when located in different countries. Conversely, the same country may produce a gem from different geological settings, such as the basaltic and metamorphic sapphires of Madagascar. The importance of origin has led to the development of new technologies e.g. the application of DNA-based nanoparticles to emerald, applied at the mine, which can survive the cutting, polishing and oiling, so the finished gemstone can be traced back to its exact place of mining.

To ensure proper disclosure and prevent use of misleading terminology, government services exist to protect the consumer, such as the UK Trading Standards and the US Federal Trade Commission. These governing bodies work with other organizations including the National Association of Jewellers in the UK. CIBJO, The World Jewellery Confederation, is an international organisation that represents the entire industry from mine to market. Its membership includes national jewellery trade organisations from more than 40 countries. To align worldwide industry standards, CIBJO developed the Blue Books, which define standards for nomenclature, grading and disclosure of treatments and synthetics. These books are devoted to diamond, coloured gemstones, pearl, coral, precious metals, gemmological labs and responsible sourcing.

There are regulations on the export and trade of some gem materials. The best-known example is that of 'conflict diamonds', brought into the spotlight in the 2000s. These are diamonds whose sale is known to fund terrorism or armed conflict (those that do not are known as conflict-free diamonds).

In 2002 the United Nations and major diamond-producing countries created the Kimberley Process, an international certification scheme to regulate the trade of rough diamonds to prevent conflict diamonds being sold on the gem market. From 2003 all rough, non-faceted diamonds (including sawn and cleaved) must receive a Kimberley Process certificate in order to cross any international border, to certify the rough was not sold to fund a conflict. Since this time the majority of diamonds traded have been certified conflict-free. Some countries, however, are not currently participating or are banned from international trade. It is critical to keep up to date with the Kimberley Process to ensure that any import or export of uncut diamonds is legal (www.KimberleyProcess.com).

CITES is the Convention on International Trade in Endangered Species of Wild Fauna and Flora. This is an international agreement between governments to protect species at risk of extinction by regulating their trade through legislation, enforcement and public awareness. CITES protects over 38,000 species of animals and plants, whether traded as live specimens or their products, including ivory, coral, tortoiseshell and some pearls. The species are listed in three appendices that are updated every two years – Appendix I includes species threatened with extinction whose trade is prohibited, Appendix II includes species not presently threatened with extinction, but which could become so if trade is not controlled, and Appendix III includes species protected in at least one country. www.Cites. org provides information on CITES, control lists of species and their status, and guides to their identification. As the species, rules and guidelines change regularly it is vital to keep up to date

These many governing organisations, rules and guidelines allow consumers to make confident informed decisions when purchasing a gemstone.

▼ Workers at the gem mines near Ratnapura, Sri Lanka.

Properties

Gems have characteristic properties by which they are defined and identified. These are directly related to their chemical composition and internal structure, influencing everything from appearance and fashioning, to use and care. Most gems are minerals, crystalline solids with a set composition and repeating atomic structure, and this unique combination dictates their physical and optical properties. However, it is important to note that minerals are not perfect. They contain impurities (elements of similar atomic size replacing essential elements in the crystal structure), have missing atoms (leaving a hole or vacancy in the structure), or the crystal structure itself is distorted. These atomic-scale imperfections can have a big impact on a gem's properties, for instance altering the colour. The study of gem properties has wide application. By understanding the cause of colour, treatments can be devised to modify that colour, and by determining the natural formation process, the process can be replicated in a laboratory to create synthetics for use as gemstones and in industry.

Table of crystal structures and symmetries

Crystal system of symmetry	Image	Structure	Crystal form	Example minerals
Isometric (also known as cubic)		Repeats the same in three directions, at 90 degrees	Cubes, octahedrons, dodecahedrons	Diamond, garnet, spinel
Tetragonal		Repeats the same in two directions, different in the third direction, at 90 degrees	Square prisms or pyramids	Zircon, scapolite
Trigonal		Repeats the same in three directions, at 120 degrees, with three-fold symmetry, and differently in the fourth direction, at 90 degrees	Three- or sixed-sided prisms, rhombohedra, scalenohedral	Quartz, corundum, tourmaline
Hexagonal		Repeats the same in three directions, at 120 degrees, with six-fold symmetry, differently in the fourth direction, at 90 degrees	Six-sided prisms	Beryl, apatite
Orthorhombic		Repeats differently in three directions, at 90 degrees	Rectangular or lozenge-shaped prism, double pyramids	Topaz, peridot, chrysoberyl
Monoclinic		Repeats differently in three directions, with two at 90 degrees, and one inclined	Parallelogram-shaped prisms	Spodumene, feldspar, diopside
Triclinic		Repeats differently in three directions, all inclined	Shapes with paired crystal faces	Feldspar, rhodonite, kyanite

CRYSTALS AND CRYSTAL STRUCTURE

Minerals are grouped into seven systems of symmetry, dependent on how their internal structure repeats. The direction of repeat is indicated by a crystallographic axis. The isometric system is the most symmetrical, with the same structure repeating in three directions at 90 degrees. Other systems have lower symmetry, repeating differently in at least one direction through the crystal.

This ordered internal structure is reflected in a mineral's external crystal shape. When the shape, or habit, is dominated by a prism it is called prismatic. This is bounded at the end by a termination such as a flat surface or pyramid. Other crystal habits are massive (lacking a defined crystal form or shape), botryoidal (grape-like), bladed, and acicular (needle-like). Aggregates of small, terminated crystals are known as druse, for example amethyst crystals lining a geode. Crystal faces may be smooth, or have fine parallel lines orientated parallel or perpendicular to the length, known as striations.

Twinning is a change in the orientation of the crystal structure, forming two or more symmetrically intergrown crystals. The surface along which they meet is the twin plane. Twinning may be simple, with mirror image crystals meeting at the twin plane (contact twin), or two crystals appearing to intergrow symmetrically (penetration twin). Polysynthetic twinning is multiple twins in parallel alignment. This is seen internally as repeated lamellar twin planes, or externally as striations on a crystal face. Polysynthetic twinning is common in corundum and feldspars. Repeated twinning may also be cyclic in a radial pattern, as seen in chrysoberyl.

Polymorphism (meaning many forms) is when a specific chemical composition occurs in more than one crystal structure, forming distinct minerals. Each mineral is stable under different temperature and pressure conditions, such as the Al_2SiO_5 polymorphs andalusite, kyanite, and sillimanite.

Pseudomorphism (meaning false forms) is when a mineral replaces another but retains its crystal form. This occurs by a mineral altering to another of similar composition, such as malachite after azurite, or replacing an entirely new structure, such as opal pseudomorphing a fossil.

Directional properties

As minerals have a directional internal structure, they have directional properties. This means the physical and optical properties can vary in different directions determined by the crystal structure. All minerals have directional physical properties, however they do not all have directional optical properties. Isometric minerals have the same crystal structure in all directions, therefore their optical properties (determined by their interaction with light) are the same in all directions, known as optically Isotropic. Minerals from all other crystal systems are optically anisotropic, having directional optical properties. Some gem materials, such as glass and opal, lack a regularly repeating crystal structure, and are known as amorphous. These gems have no directional physical or optical properties and are optically isotropic. The recognition of directional properties, or lack of, is critical to the study and identification of gems.

PHYSICAL PROPERTIES

Weight and specific gravity

The weight of a gemstone is measured in carats (ct). One carat equals 0.2 g (0.007 oz). Weight should be stated to two decimal places, and only rounded up if the third decimal is a 9, for example 0.998 = 0.99 ct but 0.999 = 1.00 ct. Many gemstones are sold by the carat, priced on a per-carat basis. The price normally jumps at set weights such as 1.00 ct, with larger stones more valued due to their

▼ Zircon (left) and tourmaline (right) faceted gemstones of the same weight, but due to zircon's higher density, it is smaller in size.

relative rarity. The size of a gemstone depends on its weight, but also its density. The density varies with composition by how heavy, and densely packed, the constituent elements are. The density of a gem is measured as specific gravity, that is the ratio of the weight of the gem to the weight of an equal volume of water. A 1.00 ct gemstone will be larger for a gem of low specific gravity than one of high specific gravity.

Durability

A gem's durability is measured by three things: hardness, toughness and stability. Durability influences how a gem is fashioned, set into jewellery and cared for. Durability also affects where a gem is found. The more durable the gem, the further it can be transported from the primary source to a secondary deposit.

Hardness

Hardness is the degree to which a gem resists scratching. It is measured using the Mohs scale of hardness. This relative scale of 1 to 10 is based on the ability of a material to scratch another of the same or lower hardness, with diamond, the hardest known natural substance, at 10. The scale, however, is not linear as diamond is several times harder than corundum at 9. Hardness can be tested by carefully scratching a gem with points of increasing hardness, however this destructive test is not recommended.

Quartz represents 7 on the Mohs scale. Gems below 7 are considered low hardness, and are less suited to everyday wear. This is in part due to the prevalence of quartz, which is found in dust, so wiping gems of lower hardness may scratch them. Some gem minerals, like kyanite, have directional hardness, varying when scratched in different directions across the crystal structure. Directional (or differential) hardness is important in gemstone fashioning and is the reason why diamond can be cut and polished by another diamond.

Toughness

Toughness is the ability to resist breaking or chipping. It is a combination of fracture and

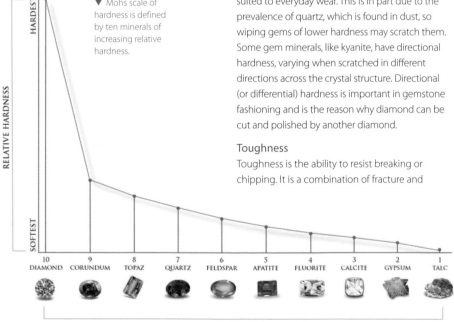

▼ Mohs scale of hardness is defined by ten minerals of increasing relative hardness.

RELATIVE HARDNESS — HARDEST / SOFTEST

| 10 DIAMOND | 9 CORUNDUM | 8 TOPAZ | 7 QUARTZ | 6 FELDSPAR | 5 APATITE | 4 FLUORITE | 3 CALCITE | 2 GYPSUM | 1 TALC |

MOHS RANKING

▶ The perfect cleavage of topaz (above) and typical conchoidal fracture of quartz (below).

may fracture gems like quartz or tanzanite. Some gems, such as pearl, react with chemicals and are damaged by perfume, hairspray and even weak acids including lemon juice or vinegar. As a rule of thumb, when worn these gems should be 'put on last and taken off first'.

OPTICAL PROPERTIES

Optical properties are defined by how a gemstone interacts with light, determining its colour, lustre and sparkle, and influencing how it is fashioned. Light is a form of energy that can be thought of as a wave, with a wavelength so small it is measured in nanometres. Light is a part of the electromagnetic spectrum, the continuous energy range from radio waves (wavelengths >100 km, >62 miles) to gamma rays (<0.01 nm). The shorter the wavelength, the higher the energy and frequency. The narrow portion of the electromagnetic spectrum that humans see is called visible light, encompassing red (~700 nm) to violet (~400 nm). Daylight consists of all wavelengths of visible light, therefore what is considered white light is actually a mix of all colours. Black is the absence of light.

cleavage. Fracture is non-directional breaking of a gem, occurring in crystalline and amorphous materials. Fracture may be distinctive, and is described by its surface, such as uneven, hackly or splintery. Conchoidal fracture is shell-like in shape, typical of glass and quartz.. Cleavage is the tendency to break (cleave) preferentially in certain directions through the crystal structure, along planes of weaker atomic bonding. Minerals have characteristic cleavage in certain directions, determined by their crystal structure. Cleavage is classed from poor (vaguely flat) to perfect (flat). Some gems, such as corundum, have no cleavage while others have multiple cleavage directions, such as octahedral cleavage in fluorite and diamond. Basal cleavage, well known in topaz, is orientated perpendicular to the crystal's length. Cleavage is a directional property and does not occur in amorphous gems. Parting, similar to cleavage, is a tendency to split along planes due to weakness but is caused by twin planes or aligned inclusions.

Polarization

Light waves travelling in the same direction vibrate in all planes at 90 degrees to the direction of travel.

Stability

Stability is a gem's resistance to light, humidity, heat and chemicals. Light may alter or fade colour, for instance in topaz. Changes in humidity can dry out opal, causing it to craze. Sudden changes in temperature can trigger thermal shock, which

▼ The electromagnetic spectrum. The energies which the human eye can see is called visible light.

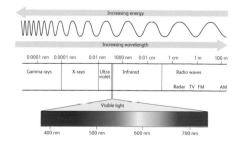

Increasing energy					
Increasing wavelength					
0.0001 nm	0.0001 nm	0.01 nm	1000 nm	0.01 cm	1 cm	1 m	100 m
Gamma rays	X-rays	Ultra violet	Infrared	Radio waves	
				Radar TV FM AM	
Visible light					
400 nm	500 nm	600 nm	700 nm		

When light waves vibrate in the same plane they are said to be plane polarized. This is a key concept in how light interacts with gemstones, and in the use of polarizing filters, which only transmit light vibrating in one plane. Crossed polarization is when two filters are placed one over the other at 90 degrees, so no light is transmitted.

Transparency

Transparency is measured by how easily light is transmitted through a gem. Transparent gems allow all light to transmit, so can be seen through. Translucent gems allow some light to pass, while no light is transmitted in opaque gems. Inclusions, fissures and aggregates of crystals (forming lapis lazuli and chalcedony) block the path of light, reducing transparency.

Clarity is a grading criteria based on the number and appearance of inclusions, regardless of transparency. Clarity grades are well known for diamonds, but as coloured stones vary greatly, their clarity grades are not standardized. Some gems, like aquamarine, are expected to be flawless while others, such as emerald, are rarely without flaws, but due to their desirable colour lower clarity is deemed acceptable. Eye clean describes a transparent gem in which no inclusions can be seen by the unaided eye.

Reflection and lustre

Reflection occurs when light falling on a surface is returned. Reflection from outer surfaces gives gemstones their lustre, and from inner surfaces creates brilliance and optical effects. Lustre is the amount and quality of reflected light. As a surface property, it is not affected by colour or transparency, however, as harder gems take a better polish they tend to have higher lustre. Opaque black diamond is cut and polished to showcase its very high lustre. Minerals have a distinctive lustre, which can be seen on crystal faces, and is maximized through polishing. Lustre is described as dull (with little reflection), resinous (amber), waxy (turquoise), silky (tiger's eye), pearly (pearl), greasy (jade) and oily (peridot). Vitreous is a glassy lustre,

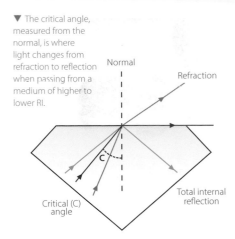

▼ The critical angle, measured from the normal, is where light changes from refraction to reflection when passing from a medium of higher to lower RI.

Critical (C) angle

seen in many gems such as quartz, and adamantine describes the super bright reflection of diamond.

Refraction and refractive index (RI)

Refraction occurs when light passes into a gemstone. As light travels from one medium (air) to another (gem) of different optical density it changes speed. If it passes through the interface obliquely, it also changes direction. This bending is known as refraction. Refraction is demonstrated when viewing a straw in water. The straw appears to bend, caused by light refracting at the water/air interface.

The optical density of a material determines how fast light can travel. A gem's refractive index (RI) is the measure of this optical density. It is calculated as the ratio of the speed of light in a vacuum to the speed of light in the gem. Considering light travels at 300,000 km/s (186,411 miles/s) in a vacuum, its speed in air is about the same. Diamond has an RI of approximately 2.4, meaning light travels through diamond 2.4 times slower than air. As the RI is constant for each gem material, it is an important identifying property, and is measured on the refractometer. The higher the density, the higher the RI, the slower light travels and the more it is refracted or bent by the gem. Gems with a high RI have high lustre and tend to create more brilliant gemstones.

▲ The strong double refraction of zircon is easily seen by the doubling of the back facets when viewed through the crown.

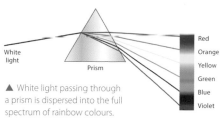

▲ White light passing through a prism is dispersed into the full spectrum of rainbow colours.

▲ Brilliant cut gemstones showing increasing dispersion and fire (left to right) fluorite, synthetic sapphire, diamond, cubic zirconia, strontium titanate.

Critical angle

When light falls on a surface it is either refracted or reflected depending on its angle of incidence. This angle, and the resultant angle of refraction and reflection, is measured from the 'normal', an imaginary line perpendicular to the surface. Light striking the internal surface of a gemstone will change from being refracted to reflected at a critical angle determined by the RI of the gem – the higher the RI, the smaller the critical angle. The concept of critical angle and total internal reflection is used when cutting gemstones to produce maximum brilliance.

Double refraction and birefringence

When light passes through an optically anisotropic gem, it is split into two components called rays. These rays are plane polarized, vibrating at 90 degrees to each other, travelling at slightly different speeds. Each ray refracts differently with a distinct RI and takes a slightly different path. This is known as double refraction or birefringence. Birefringence is measured as the maximum difference between the two RIs. In gem materials with high birefringence, the doubling effect is so strong it can be seen by eye. Calcite is well known for this, with a cross seen through a clear crystal appearing doubled. Double refraction can be observed in faceted gemstones by a doubling of the back (pavilion) facets and any inclusions, when viewed through the top (crown) facets, as observed in zircon and peridot.

Dispersion and fire

Dispersion is the spreading of white light into the full spectrum, seen when light passes through a prism. This occurs because each colour wavelength travels at a slightly different speed, so is slowed and refracted a different amount, spreading into the rainbow. As the RI of a material varies with wavelength, the dispersion of a gemstone is measured as the difference between the RI for red (687 nm) and violet (431 nm) wavelengths. The greater the difference, the higher the dispersion, and the wider the rainbow seen. This effect is often described as fire. Dispersion is sometimes classified as low (<0.017 e.g. glass, quartz), moderate (0.017–0.050 e.g. corundum, and surprisingly diamond), high (0.051–0.071 e.g. demantoid garnet, cubic zirconia) and very high (>0.071 e.g. synthetic moissanite, strontium titanate).

▶ Light reflecting from fine layering in this fire agate diffracts and interferes to create iridescent colours.

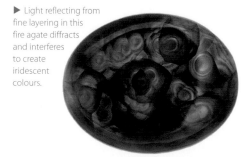

Diffraction and interference

The spreading of light waves into the spectrum also occurs when light passes through a very small hole or slit, or is reflected from tiny striations or particles. This is known as diffraction. Interestingly, red wavelengths are diffracted more than violet, the opposite to dispersion. A diffraction grating has multiple apertures in a regular arrangement. Interference occurs when two light waves travelling in the same direction interact, combining to boost the colour, or cancelling each other out. This is the cause of iridescence. When white light is diffracted and spread out, it interferes with light waves diffracted from neighbouring apertures. The interference of the different colour wavelengths creates bright spectral colours. This is the cause of opal's play of colour, labradorite's iridescence, and in part the soft 'orient' of pearl.

Brilliance

Brilliance is the brightness of a cut gemstone when viewed face up. It is the quantity and quality of the light reflected from the inside surface of the pavilion facets, back to the viewer. Transparent gems, free from inclusions (which potentially block the light) show greater brilliance. It is dependent on the RI of the gem, and its critical angle (at which light changes from refraction to reflection), the cut of the gemstone and the angle of the pavilion facets. These determine whether light entering the gemstone is totally internally reflected back to the viewer, creating a lively stone, or if it is refracted out through the pavilion, leaving it lifeless.

Scintillation

Scintillation is the twinkling effect seen as a gemstone is moved. If light reflects internally from a higher number of smaller pavilion facets, it breaks up the reflections creating contrasting areas of light and dark. These flash in and out as the stone is turned, generating sparkle.

Colour

The colour of a gem is not a physical property of the gem; rather it is the colours of light the gem reflects and/or transmits. Incident light energy interacts with the stone. Certain wavelengths are absorbed, as determined by the gem's composition and structure. The other reflected or transmitted wavelengths are seen by the viewer as a colour. If all colours of light are reflected the object is seen as white, and when all light is absorbed, it appears black. If all incident light is transmitted it appears colourless.

The absorption of a wavelength's energy is usually due to certain elements, such as iron and chromium. These elements are the cause of a gem's colour, known as chromophores. How these elements are arranged in the crystal structure influences the colour seen, for example chromium imparts both emerald's green and ruby's red. When the chromophore is an essential component, such as copper in azurite and malachite, the mineral is idiochromatic (self-coloured), while if the chromophore is an impurity as chromium is in ruby, it is allochromatic (other coloured). Defects in the crystal structure can also absorb light of particular wavelengths, known as colour centres.

▼ The absorption or reflection of different wavelengths of light determines what colour an object appears.

A red object reflects red and absorbs others colours of white light

A white object reflects all colours of white light equally

An object is seen as black if it absorbs all colours of white light

While colour is arguably the most important aspect of a gem's beauty, it is subjective. Everyone 'sees' colour differently. Different types of illumination contain different wavelengths, which dramatically influence the perceived colour. Colour is described by the hue (colour), tone (relative lightness or darkness) and saturation (the intensity). Some gems are a combination of hues, such as yellowish green, where green is the primary hue and yellow is secondary. Coloured gemstones such as fancy colour diamond are evaluated by these qualities, however there is no standardized grading system.

Pleochroism

Pleochroism means many coloured. As light passes through optically anisotropic gems, its interaction changes depending on its vibration direction within the crystal structure. This means the wavelengths absorbed may change, causing the gemstone to appear as different colours when viewed from different directions. Additionally, as the light waves entering the gem are also split into two rays vibrating at 90 degrees, each ray interacts differently, so can be a different colour. This is seen with a dichroscope. The eye however combines the two colours, only 'seeing' the resultant colour in each direction. Uniaxial minerals can show two pleochroic colours, called dichroism. These are minerals in the tetragonal, trigonal and hexagonal crystal systems, whose symmetry allows light to vibrate in two directions/planes through the crystal structure. Biaxial minerals can show three colours (although only two in any one direction) called trichroism. These minerals are in the orthorhombic, monoclinic or triclinic crystal systems, with symmetry allowing vibration in three directions.

Pleochroism is classed as strong, distinct, moderate, weak and none. Colourless and optically isotropic gems (isometric and amorphous) do not show pleochroism. Pleochroism has a huge influence in fashioning gemstones, with cutters orientating the rough to obtain the best pleochroic colour, or cleverly show more than one colour, face up.

▶ An iolite in a swivel ring, cut as a cuboid to show a pleochroic colour through each face, showing dark blue when viewed vertically and straw yellow horizontally.

▼ As light passes through this aquamarine, the dichroscope reveals the pleochroic colour of each ray (dark and light blue), however the eye combines the colours so the crystal is seen as medium blue.

Colour change effect

Colour change gemstones are those that appear to change colour when viewed under different lighting. This is known as the alexandrite effect, due to the remarkable red to green change seen in the alexandrite variety of chrysoberyl. This effect is different to, and independent from, pleochroism, although the colour change may be strongest when viewed in a direction of pleochroism. Colour change occurs when a gem transmits light in two parts of the visible spectrum. Alexandrite transmits wavelengths in both the red and blue-green. When viewed in daylight, which has a high proportion of blue and green wavelengths, it appears green, aided by the eye being more sensitive to green light. When viewed in tungsten or candlelight, which contain more red than blue wavelengths, it is enough to tip the balance and the alexandrite is perceived as red. Other gems including corundum, spinel, garnet

▲ The 'Murchinson Snuff Box' set with 16 large diamonds, seen under normal light (left) and UV light (right), revealing the differing fluorescent response of the diamonds.

and diaspore exhibit this phenomenon, and those with a strong colour change are highly valued.

Luminescence

Fluorescence, a type of photoluminescence, is the phenomenon whereby a material temporarily absorbs light energy, re-emitting it at a lower energy (longer wavelength). This effect is seen when a gem absorbs UV light, releasing the energy as visible light and appearing to glow. The colour of the fluorescence varies from the gem's normal colour, with shortwave (254 nm) and longwave (365 nm) UV light eliciting different responses. Certain elements such as chromium, rare earth elements and uranium cause fluorescence, while the presence of iron quenches the response. As certain responses are characteristic, fluorescence is a useful tool for identifying gems, indicating treatments (e.g. heated sapphires or fracture fillings), and distinguishing synthetics (e.g. diamonds). In some gems the fluorescence reaction is so strong that the ultraviolet in daylight is enough to cause a glow, seen in ruby, fluorite, blue amber and some hyalite opal. Fluorescence in other parts of the electromagnetic spectrum is used for gem analysis, including the measurement of emitted X-ray fluorescence.

Phosphorescence occurs when a gem continues emitting light after the UV light source has been removed, seen in some opals and blue diamonds. The emission of light can be generated by methods other than absorbing light energy. Triboluminescence is the emission of light after being struck or rubbed and thermoluminescence occurs when a material is heated – as seen in fluorite.

INCLUSIONS

Inclusions are foreign materials or irregularities incorporated into a gem. They may be solid (such as a mineral crystal), liquid or gas trapped in variously shaped cavities, or a combination of solid, liquid and gas in two or three phases. Other features classed as inclusions are colour zoning (caused by changes in the available elements during crystallization), twin planes, fractures and stress cracks, and incipient cleavage (where a gem has begun to split along a cleavage plane forming a mirror-like internal surface). When a fissure partially heals, it leaves small inclusions following the plane of the fissure of trapped fluids or negative crystals (voids in the crystallographic form of the host gem). These form feathers or fingerprints, named for their appearance.

Protogenetic inclusions are older than the gem, incorporated as the gem grew around them. These are commonly crystals of other minerals, such as the wonderful inclusions in quartz. Syngenetic

▲ Silk inclusions, exsolved after the sapphire formed, are orientated in the sapphire's trigonal symmetry. As the silk is intact, it indicates this gem is unheated. FOV 2.01 mm.

inclusions formed at the same time as the host, and are often crystallographically orientated within it. This includes zoning, multiphase (solid, liquid, gas combination) inclusions, negative crystals and twinning. Epigenetic inclusions form after the gem has stopped growing. This may be exsolved crystals such as silk in corundum, and feathers and fingerprints of rehealed fractures.

Inclusions are often considered detrimental to the value of a stone, particularly for gems such as diamond, which is valued for its clarity. Some gems, however, are prized for their inclusions, such as the plants and insects preserved in amber, and the inclusions that cause optical phenomena such as chatoyancy and asterism. Inclusions are also incredibly important in the study of gemstones, and the history and geology of Earth, providing information on the conditions of gemstone formation. Inclusions can be used to indicate provenance, and reveal if the mineral is natural, heated or synthetic.

OPTICAL PHENOMENA

Phenomenal gems are those that exhibit an optical effect beyond their colour. These are caused by inclusions or internal structures that interact with light. Chatoyancy, or cat's eye effect, is caused by light reflecting from multiple parallel inclusions. These are elongate needle- or hair-like mineral inclusions, or tubular cavities, which cause a silky reflection. The

▶ The glittering green aventurescence in this quartz carving is caused by tiny included fuchsite mica flakes.

cat's eye effect is revealed when the gem is cut en cabochon, with the base parallel to the plane of the inclusions, or in a sphere. The reflection forms a line or band of light perpendicular to the needles, which moves over the stone. The finest cat's eyes are seen in chrysoberyl, with a sharp white band on translucent honey-coloured gems. Many other gems show this phenomenon, albeit more diffuse, including tourmaline, beryl and quartz. Chatoyancy is best seen by a single pinpoint light.

Asterism (from the Greek 'aster' meaning star) is a similar reflection effect, where the parallel inclusions are orientated in two or more directions in the same plane. Each set reflects a band of light that then intersect to create a four-, six- or 12-rayed star. To reveal the effect, the gem must be fashioned en cabochon, correctly orientated with the horizontal plane parallel to the inclusion plane, or in a sphere. The crystal structure determines the orientation of the inclusions and therefore the star. For example, corundum is trigonal and commonly contains fine needles orientated in three directions at 60 degrees. This creates three intersecting bands forming a six-rayed star. Star diopside is monoclinic, containing two sets of inclusions. This causes two light bands intersecting at an oblique angle in a distinctive four-rayed star. Asterism is similarly best seen with a single pinpoint light. Interestingly, rose quartz exhibits a star effect by both transmitted light and reflected light.

▶ A rare 12-rayed star sapphire. The gold arms are reflections from rutile inclusions orientated in three directions, and the silver arm reflections from hematite inclusions also in three directions but offset by 30 degrees, creating a 12-rayed star. Hexagonal growth zoning is also visible.

Aventurescence is a metallic glittery effect caused by inclusions of tiny flakes or platelets. These have the same orientation, reflecting light to create the shimmery effect. This phenomenon is seen in aventurine quartz, 'bloodshot' iolite and sunstone feldspar. The name is derived from the Italian 'a ventura' meaning by chance, following the reputedly unexpected creation in the early 1600s of a glittering copper-included glass called goldstone or aventurine glass.

Adularescence is a milky or bluish reflection of light seen beneath the surface of a gem. This sheen is caused by the scattering of light from tiny particles or layers inside the stone. When these submicroscopic features are smaller than the wavelengths of visible light, blue light is more strongly scattered and the sheen appears bluish. It is seen in moonstone and other feldspars including

▶ This sapphire contains fine inclusions known as silk which scatter the light creating a milky bluish sheen.

adularia from which it is named. Opalescence is the form of adularescence seen in opal.

Iridescence (from the Greek 'iris' meaning rainbow) is the changing rainbow colours caused by the interference of light. Light is reflected from one or more transparent thin films where the layer has a different refractive index to its surroundings, e.g. coatings, air-filled cracks, internal layering of twinning or lamellar structures. Light waves are reflected from the top and bottom of each layer, travelling side by side and interfering to create rainbow colours. Thin film interference is seen in rainbow garnet, ammonite and fire agate, and is the cause of iridescence in thin film coatings on topaz and quartz.

Labradorescence is a type of thin film interference, seen in its namesake labradorite feldspar. Labradorite contains submicroscopic internal layering (smaller than the wavelengths of visible light) of alternating composition, RI and thickness. The cause of the phenomenon is complex with light reflecting, refracting and diffracting from the layers, interfering to create metallic spectral colours.

Play of colour is the colourful iridescent flashes seen in precious opal. It is caused by the diffraction and interference of light. Opal consists of tiny spheres of silica stacked in a regular array that act like a diffraction grating. The gaps between the spheres are close to the wavelengths of visible light, and small variations in size determine the colour and size of the flashes.

▼ The bright spectral colours of iridescence, caused by thin film interference from a fracture, reveal the 'tiger stripe' inclusions in this amethyst gemstone. FOV 14.35 mm.

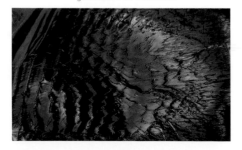

▶ A black opal cabochon set in a ring with spectacular play of colour caused by the diffraction and interference of light.

Cutting and fashioning

Most gem materials are cut and shaped to reveal or enhance their beauty, creating sparkle and highlighting colour and lustre. The way in which they are fashioned depends on many factors – their appearance, durability, optical properties, the presence of inclusions and optical phenomena. The art of gem cutting is separate for diamond (known as diamond cutting), and coloured stones (called lapidary). Both are a balancing act between creating a beautiful gemstone with optimum colour and brilliance, and retaining the most weight, with the ultimate goal to obtain the highest value.

For thousands of years, natural materials have been fashioned into weapons, tools, ceremonial objects and for adornment. One of the earliest techniques was simply to create a hole to make pendants and beads. Techniques developed, such as slowly shaping softer material through abrasion by a harder material (a process now known as bruting), and utilizing cleavage to split gems into smaller workable sizes. The art of engraving and carving, called glyptic, goes back thousands of years as illustrated by the scarab amulets of Ancient Egyptians. Colourful, easy-to-work gems were favoured, such as lapis lazuli, amber, turquoise and malachite, but the carving of harder jade and quartz varieties (agate, carnelian, amethyst) was also mastered.

One of the first cutting styles was the domed cabochon. The technique of faceting gemstones only began in the 1400s, with the invention of a flat horizontal polishing wheel known as a scaife or lap. Abrasives of varying hardness were mixed with water or oil to lubricate and prevent the stone overheating. This revolutionized the way gemstones were cut, allowing flat symmetrical facets at specific angles. Techniques initially focused on diamond, developing ways to improve sparkle and fire, and these were in turn applied to coloured stones.

The history of faceting gemstones has changed dramatically over the past 500 years. The understanding of mineralogy, crystal structure and symmetry from the late 1700s improved cutting through awareness of optical properties. The mechanization of gem cutting by the invention of the bruting machine in 1870s, and the motorized saw around 1900, brought cutting into the modern era. Today accurate techniques utilizing virtual 3D modelling, laser cutters and other equipment produce gemstones of the highest yield with ideal proportions and symmetry. Machines cut calibrated gemstones in standardized sizes for use in jewellery designs.

TYPES OF CUT

Freeform
The simplest method of fashioning is to simply polish the surface of the gem, revealing the lustre, colours and textures, and following any shape in the rough.

Cabochon
The cabochon is one of the oldest, simplest and most popular polished forms. Opaque and translucent gemstones are commonly fashioned as cabochons, as well as transparent gems of lower clarity. This smoothly domed form ranges from almost flat to steep. The rear may be polished, and is flat to convex, particularly in rare or high-value gems, to increase the weight, and therefore price. Occasionally, very dark gems are cut with a concave back to allow more light to pass through and lighten the colour, commonly used for dark red garnets. The rounded shape lacks corners, offering fragile gems protection against damage. Oval and circular outlines are prevalent. High-quality cabochons have good symmetry on the dome, base and outline.

Cabochons promote a gem's colour, texture, and lustre, and for translucent gems such as jade, make the colour glow. This style also reveals optical phenomena. Chatoyancy (cat's eye) and asterism (star effect) are caused by reflections from numerous parallel inclusions. The cabochon must be orientated with its base parallel to the plane of the inclusions. Very fine needle-like inclusions result in a translucent/transparent stone with a sharp eye or star, while thicker or platy inclusions create a diffuse effect in an opaque stone. Medium to high uniform domes are most effective and the finest quality have a central sharp line or star with straight even arms. Natural star stones are often slightly asymmetric in shape as determined by the rough, and are cut deep to retain more inclusions, and weight. Synthetic star stones can be created, commonly for ruby and sapphire. These exhibit a very sharp star in an often perfectly symmetrical cabochon. A poorer polish and flat back may also indicate a synthetic stone.

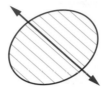

▲ The orientation of the inclusions within a cabochon to reveal the cat's eye effect.

▲ This apatite is more opaque with a diffuse cat's eye due to larger needle-like inclusions.

▲ Chrysoberyl forms the finest cat's eye with a transparent body and sharp bright line.

◀ This superb mocha stone cabochon was carefully polished to bring the dendritic inclusions to the surface, creating the image of a tree, and achieving a symmetrical shape.

▼ Almandine cabochons are commonly fashioned with a concave back to allow more light through and lighten the colour. These are commonly referred to as carbuncles.

Cabochons intensify optical phenomena such as play of colour (the bright spectral flashes seen in opal) and adularescence (the milky sheen of moonstone).

Beads and spheres

Beads and spheres are commonly fashioned from opaque or heavily included gems, but also from transparent material. Beads come in many shapes and may be rounded, baroque (tumbled stones or polished chips), or faceted to utilize lustre. The rounded shapes offer protection for more fragile gems. Spheres, created for both adornment and decoration, highlight attractive colours and textures, and can also reveal chatoyancy and asterism.

Faceted gemstones

Faceting is the art of cutting and polishing flat surfaces, known as facets, in a symmetrical arrangement. Predominantly used for transparent gems, the facets create the sparkle, reflecting light around inside the gem and returning it back to the viewer. Different faceting styles are chosen for diamond and coloured stones. Diamond, prized for its lack of colour, is cut to maximize brilliance, fire and scintillation, achieved by the brilliant cut. Pale coloured gems with a high refractive index (RI), such as chrysoberyl and garnet, are cut for brilliance, while those with high dispersion, like zircon, titanite and sphalerite, are cut to show fire, although this is masked by darker colours. Coloured stones are cut in a variety of styles, such as mixed or step cuts, which emphasize and even out colour. The style and quality of cut have a huge impact on the perceived colour of the finished gem. Opaque gems are sometimes faceted to promote their high lustre.

Light reflecting from the crown facets, those above the girdle, provides lustre. Light entering the gem through the table facet which is then totally internally reflected by the pavilion facets, those below the girdle, creates brilliance. The inclined surfaces of the facets also disperse the reflected light into spectra, creating fire.

A gem's brilliance is determined by the gem's RI and the angle of the pavilion facets. The pavilion determines the angle at which the light hits the facets (angle of incidence), and the RI determines the gem's critical angle, beyond which the light is reflected rather than refracted.

If the stone is shallow cut, so the incident angle of the light is less than the critical angle (i.e. closer to the normal), light is refracted. It passes out the back of the stone creating a 'window', and the gem lacks brilliance. If the stone is well cut so that the incident angle is greater than the critical angle (i.e. further from the normal), light is totally internally reflected, exiting back to the viewer. This creates a lively stone, and in coloured gemstones intensifies colour. If the stone is cut too deep, the light is still internally reflected, but exits via the sides, creating dark areas known as extinction.

By knowing the RI and critical angle, a gem can be designed with pavilion facets at the perfect angle, so that all light entering through the table is totally internally reflected back to the viewer, maximizing brilliance. Gems with a high RI have smaller critical angles, and are more likely to show brilliance. Gemstones with a low RI must have a particularly steep pavilion to achieve total internal reflection, so are generally cut for colour rather than brilliance. Scintillation is achieved by using many smaller facets on the pavilion. This breaks up the internal reflections so the gemstone will twinkle as it is turned.

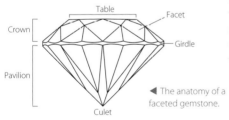

◀ The anatomy of a faceted gemstone.

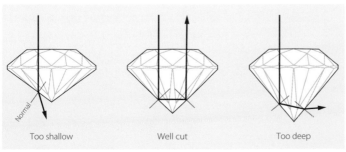

Too shallow   Well cut   Too deep

◀ Illustration of the path of light through a faceted gemstone and the effect of the pavilion angle. In well cut stones the light is reflected back to the viewer. With shallow pavilions light exits the back creating a window, and if too deep light exits through the side creating extinction.

FACETED CUTS

The pattern of the facets and the shape of the cutting style are designed to make the most of a gem's properties. Displayed here is a selection of frequently seen faceted styles.

Brilliant

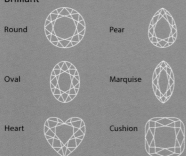

Round   Pear

Oval   Marquise

Heart   Cushion

Step

Step cuts have rectangular stepped facets.

Baguette   Emerald

Mixed

Mixed cuts combine brilliant cut and step cut styles.

Scissor

Scissor cuts have facets arranged in a criss-cross pattern.

Fancy

Fancy cuts have unusual shapes and facet arrangements.

Fantasy

Fantasy cuts have unique facet patterns and shapes including grooves and concave facets.

The quality of a faceted gemstone is determined by the proportions, symmetry and polish. High-quality cuts have brilliance without windowing or extinction, a pleasing internal pattern of contrasting light and dark, and a symmetrical shape with regular, aligned facets, sharp facet edges, and mirror-flat polish.

Brilliant cut

The brilliant cut was designed in 1919 by Marcel Tolkowsky. The modern brilliant cut has 57 facets (or 58 with a small facet on the culet) to create a bright sparkling gem. Designed specifically for diamond's RI and dispersion, it has the exact proportions and angles to achieve total internal reflection and provide maximum brilliance and fire. Brilliant cuts are fashioned in a range of outlines including round, oval, heart, pear (drop or pendeloque), marquise (navette or boat-shaped), and cushion (squarish with rounded corners). Round is so commonly used for diamond it is known simply as a brilliant cut.

Step cut

The step cut (also trap cut) has a series of elongate facets around a large table. It is fashioned in many outlines, usually rectangular or square. Long rectangles are known as baguettes, and rectangles or squares with truncated corners are called emerald cuts. This provides more protection for brittle stones, hence is used extensively for emerald. The step cut emphasizes colour, but does not hide inclusions well, so is better suited for higher clarity gemstones. As it is not concerned with brilliance or fire, it allows deeper or shallower pavilions to obtain best colour. This simple style also highlights colour zoning.

Mixed cut

The mixed cut is usually a brilliant cut crown with a step cut pavilion. This style is often used for coloured stones, helping to even out hue, also allowing a deeper pavilion to enhance the colour. Oval and cushion shapes are most common.

Scissor cut

The scissor cut has triangular side facets making an x shape. This cut is often used for synthetic or imitation gemstones as it is simpler to create with fewer facets, and hides irregularities that occur due to lower quality cutting.

Fancy cuts

Fancy cuts do not use the traditional arrangement of facets, but instead have unusual combinations to create a unique internal contrasting light pattern. This style is commonly used for coloured and particoloured gems.

Fantasy cut

Fantasy cuts are artistic creations with unusual outlines, concave facets, or grooves to create new light patterns and optical illusions. They are often used for coloured gemstones, and utilize colour zoning.

Engravings and carvings

Gems have been engraved for many years, historically used as seals by pressing into wax. Carvings use harder, more durable gems that lack cleavage, including quartz and chalcedony varieties, tourmaline, corundum, jade and aquamarine, thus reducing the risk of breaking or chipping. Other opaque decorative gems such as rhodonite and sodalite, and the softer turquoise, amber and jet, are also used.

Cameos and intaglios are carved or engraved gemstones depicting a portrait or scene. A cameo is raised above the background, while intaglios have the image carved into the gemstone. Gem carvers use layered or banded material

such as shell or agate, and cleverly incorporate the different colours into the design. These were highly valued in ancient times, regaining popularity in the Renaissance.

Inlays and mosaics

Inlays are thin slices of gem material set flush into a surface. Opaque, brightly coloured gems such as malachite, tiger's eye and lapis lazuli are used in attractive patterns, or to form a mosaic. Inlays are used in furniture such as inlaid tabletops, decorative items and jewellery including bangles.

▲ A cameo cleverly using a pink agate layer on amethyst druse, which formed the interior of a geode, as seen by the curvature. Probably from Brazil, and engraved by Wilhelm Schmidt.

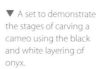

▼ A set to demonstrate the stages of carving a cameo using the black and white layering of onyx.

▲ An intaglio with the image engraved into the surface of this green tourmaline gem. Engraved by Wilhelm Schmidt.

CUTTING TECHNIQUES

The cutting of diamond and coloured stones is considered two different processes, although the steps are similar. The basic steps to cut a gemstone are:

- Sawing or cobbing (knocking off brittle or fractured sections) to remove unwanted material, or split large rough into several pieces. Sawing uses a thin circular blade with diamond dust on the outer edge. Diamonds may also be cleaved into smaller pieces, or cut by lasers.

- Preforming by grinding (bruting) and sanding the rough into the desired shape, called a preform. Preforming is done using a lap with coarse then finer abrasives. This is the most important step as it orientates and shapes the gemstone within the rough, determining size, shape, symmetry and proportions. Cabochons and faceted gemstones are then attached (dopped) to a wooden or metal dopstick with wax, epoxy or glue.

- The grinding of facets is the next step for faceted gemstones. The dopstick is inserted into a handpiece, which holds it against a lap. This allows precise positioning of the stone both vertically and rotationally. The facets are ground, sanded then polished by the rotating lap using a lubricant and coolant such as water. Cabochons are twirled to create smooth surfaces.

- The last step is polishing to remove all scratches, using progressively finer abrasives and even leather, felt or cork.

Drilling and carving

Holes are created using a small rotating rod/tube with a diamond-tipped (or other abrasive) drill bit, or an ultrasonic drill that vibrates at extremely high frequency. Gems may be partly drilled, to attach to posts, such as stud earrings, or fully drilled to create a bead. Carving may be done by hand using diamond-tipped burrs, or by a machine.

Tumbling

Gemstones may be tumbled in a rotating barrel, in water with progressively finer grades of abrasives. This creates polished surfaces and rounded edges on irregular shapes known as baroque. Tumbled stones may be drilled for use as beads.

FASHIONING

There are many factors to consider when cutting gemstones to achieve greatest beauty when viewed face up (i.e. through the top of the gem), and to retain highest yield. Transparent gemstones are usually faceted, as the unimpeded interaction of light with these flat, angled surfaces creates brilliance, fire (dispersion) and sparkle (scintillation). Colourless and pale coloured gemstones are cut to emphasize these attributes. For coloured gemstones, colour is the most important factor in determining value, so fashioning will focus on optimising it. Opaque to translucent material is polished in rounded styles, such as cabochons, to promote colours, patterns, and optical effects without concern for brilliance and fire.

▼ The stages to facet a brilliant cut diamond from an octahedral crystal, modelled in rock crystal.

When cutting the rough, the decision on the shape and orientation of the final gemstone is determined by several factors:

- Shape of the rough – this will constrain the shape and weight of the finished gemstone.

- Transparency – transparent gems are more likely to be faceted, while opaque gems are polished into cabochons, beads or decorative items.

- Inclusions – the gemstone is usually orientated to exclude detrimental inclusions or fractures during cutting, or to minimize their appearance, such as silk in corundum. Other gems, including rock crystal, are cut to highlight attractive inclusions.

- Optical phenomena – the cabochon cut reveals effects such as chatoyancy and asterism, and promotes play of colour and adularescence.

- Hardness – gems with differential hardness are harder to polish in certain directions. Softer gems will not achieve the sharp, flat facets of harder gems.

- Durability – gemstones of lower durability are fashioned in rounded shapes, or with truncated corners, such as the emerald cut, to protect against knocks.

- Cleavage – gems cannot be polished along a cleavage plane. The rough is orientated so facets are not parallel to any cleavage planes, also reducing the risk of later breakage.

- Double refraction – the rough is orientated to minimize strong double refraction, as seen in zircon, otherwise the gemstone can appear blurry from the internal doubling effect.

- Colour zoning – zones of different colours may be utilized to create attractive particoloured gemstones, seen in tourmaline and fluorite. Other gems are orientated to hide zoning, or cut asymmetrically to appear uniform in colour when viewed face up, utilized in sapphire and amethyst.

- Pleochroism – important when orientating pleochroic stones, as it has a dramatic effect on the final size and shape, especially for strongly

◀ The elongate crystals and pleochroism of tourmaline significantly influence the shape and colour of the finished gemstone.

◤ This particoloured tourmaline gemstone is cut to utilize the attractive colour zoning.

▼ The strong pleochroism of tourmaline is seen in this gem as a blue green 'bowtie' across the middle, and yellow green at the top and bottom.

pleochroic, elongated minerals like tourmaline. For dark gems, the table facet is orientated to show the lightest pleochroic colour face up, and vice versa for light coloured gemstones. Some gems are cut to exhibit a more desirable colour face up, such as the violet-blue of tanzanite and iolite. Other gemstones are cleverly cut to show flashes of different pleochroic colours at the same time, seen in diaspore, andalusite and sillimanite.

- Colour change – orientated to show the strongest colour change face up.

- Depth of colour – delicately coloured gemstones, such as spodumene and beryl varieties, are cut with a deeper pavilion to enhance colour, while darker colours are shallow cut to lighten them. The saturation of colour increases with gemstone size. This is desirable for tanzanite with the best colour seen in gems more than 5 ct, however for chrome diopside the colour becomes too dark, and smaller stones are preferred.

Treatment

A treatment is a procedure applied to a gem to improve its appearance and/or durability, other than traditional cutting and polishing. Improving the beauty, and therefore desirability, of inferior material allows rarer gems to become more widely available and affordable. For instance, heating the abundant milky white 'geuda' sapphire produces attractive transparent colours, and blue and pink topaz, rare in nature, can be created through different treatments.

▲ 'Mystic' topaz with a thin metallic coating on the gemstone's pavilion, creating iridescence. The ring setting protects this softer coating.

Treatments are applied to the rough, preformed or finished gemstone, sometimes involving multiple procedures to achieve the desired effect. Detection may be easy, for instance by observing an unnatural colour, or require advanced testing. Some treatments are undetectable. Certain treatments are stable and permanent, others can deteriorate or fade over time. Some treatments improve a gem's durability, conversely others will ultimately weaken the gemstone. As treatments can be difficult to detect, and have an impact on value, there are legal requirements for specific or general disclosure at the point of sale. Those requiring specific disclosure need verbal disclosure (prior to and/or at the time of sale) in addition to a full written disclosure. Other, often less invasive, treatments require general disclosure involving only verbal disclosure. All labelling should indicate treatments. Governing bodies, such as Trade Commissions and CIBJO, provide regulations and guidelines.

TYPES OF TREATMENTS

Surface coating

Thin coatings are applied to a gem's surface to fill imperfections, improve the lustre, colour or durability, and provide protection against skin oils. Colourless coatings can intensify colour, translucency and lustre, and improve the hardness of softer gems. These may require general or specific disclosure. Wax is routinely applied to jade, and thin coatings of wax, polymer or resin to opal, coral, pearl or turquoise. Coloured coatings are applied to cut gemstones to improve or alter hue, and require specific disclosure. This treatment is used on topaz, quartz, diamond, tourmaline and tanzanite. Metallic coatings are applied by the process of thin film deposition, where metals are deposited by vapour as a film. The coating is often applied to the pavilion so it is protected when the gem is set in jewellery. Different metals or metallic oxides are used – titanium oxide creates rainbow iridescence on 'mystic' topaz or quartz, and gold imparts the blues of 'aqua aura' quartz. Synthetic diamond is also used, grown as a thin coating by chemical vapour deposition.

Impregnation

Porous, powdery or cracked material may be permeated by colourless wax, oil, epoxy resin or acrylic. This stabilizes the gem improving hardness, toughness, clarity, lustre and enhancing colour. Most turquoise is stabilized by impregnation, as well as lapis lazuli, azurite and malachite. Jadeite jade has a porous granular texture and is commonly impregnated, particularly after bleaching which weakens its structure. Wax and oil are less permanent than polymers or resins. Impregnation requires specific disclosure.

Bleaching

Gems are bleached using chemicals to lighten, whiten or even out colour. Cultured pearls are routinely bleached with hydrogen peroxide, black coral is bleached to produce golden colours and strong acids are used on jadeite jade to remove brown staining. Bleaching is often undertaken before dyeing. This treatment can be harsh, weakening the structure of porous gems and decreasing durability. Bleaching often only requires general disclosure.

Dyeing

Any permeable gem material can be dyed, to enhance or change its colour. Dyes and stains penetrate through interconnected pores or fractures, with layers of higher porosity absorbing more dye. Dyeing can enhance the natural colour, for example in jadeite jade, turquoise and red coral, or create an unnatural colour, seen in dyed agate and some pearls. Certain materials are dyed to simulate more valuable gems such as dyed howlite to imitate turquoise. Some dyeing techniques have been used since ancient times. Opal and agate are traditionally heated in a sugar solution, then sulphuric acid, causing the sugar in the pores to convert to black carbon, darkening the gem. Pearls are treated with a silver nitrate solution that reacts on exposure to light, decomposing into metallic silver to create dove grey colours. Quartz may be heated and

quickly cooled, inducing fractures that allow the penetration of coloured dyes, known as quench crackled quartz. Dyes are often added to fracture-filling oils or polymers to simultaneously improve clarity and colour, as used in emerald and ruby. Dyeing is a common treatment and requires specific disclosure. Dyes can be detected by colour concentrations in fractures, and different fluorescent reactions. They are not always permanent, but can leak, or be dissolved by alcohol, acetone and other chemicals, so care must be taken to avoid contact. Dyes may also fade over time with exposure to sunlight.

Fracture filling

Surface-reaching fractures are filled with wax, oil, polymer/resin, or glass to improve the transparency and colour of the gemstone. By using a filler of similar refractive index (RI) to the gem, the fractures become less visible as light passes through, instead of reflecting off, the fracture interface. Treatment can occur under vacuum to draw out the air, allowing fractures to be completely filled. Most emerald, known for its flaws, is routinely fracture filled, and the oiling of emerald has occurred since ancient times. Cedarwood oil is commonly used today, as well as polymers or resins with a similar RI. Fracture filling is utilized on ruby, and to a lesser extent diamond, using glass of high lead content. Many low-quality rubies have such a high proportion of lead glass filling they are termed composite ruby. Other treated gems include tourmaline, red beryl, opal and occasionally alexandrite, peridot and tanzanite. Fillers are often coloured to strengthen the gem's hue.

Oiling improves appearance but does not alter the durability. Resin and glass can improve toughness by bonding the gemstone. As mild chemicals and acids, or heat, can damage lead glass filling, treated gemstones, especially composite ruby, require special care.

Fracture filling is not permanent. Oil can leak, dry out or decompose over time becoming less transparent

◀ Quench crackled quartz, with induced fractures containing red dye, used to imitate ruby.

or lumpy. Care must be taken to avoid heat, detergents and chemicals to prevent accidental removal of the oil, however retreatment is possible. Polymers, resins and glass are more stable, but harder to remove if they decompose or devitrify. Fracture fillings can be detected by visual inspection with a hand lens, having a different surface lustre or containing trapped bubbles. When rotating the gemstone, light reflected from the internal fissure interface gives a characteristic coloured flash. The filler may also cause colour concentrations in the fissures, or fluoresce differently to the gemstone. Treatment with colourless oil requires general disclosure, while use of resins, glass and coloured oils require specific disclosure.

Heating

Heating is one of the oldest forms of gemstone treatment. Heat alters colour, and may reduce the appearance of inclusions and fractures, improving clarity and brilliance. As numerous gem materials are routinely heated, deemed acceptable by the gem trade, it requires only general disclosure unless combined with flux healing of fissures, or diffusion. Heating allows the elements in a gem to undergo chemical changes or migrate in the crystal structure. This alters the way light interacts with the gem, thus altering its colour. Inclusions may be absorbed, and fractures healed, improving transparency. The perfect example is rutile silk in sapphires. When sapphires are heated to high temperatures the rutile (TiO_2) is dissolved, improving transparency and brilliance, freeing up the titanium required to give the gem its desirable blue colour.

Many gems undergo heat treatment including ruby, sapphire, zircon, kunzite, amethyst and carnelian. Gems react differently to heat depending on their source, due to varying impurities and inclusions. For some, such as tanzanite, the treatment is permanent and stable. Others, like kunzite, will fade with exposure to light and/or heat. Heating can also make gems such as zircon more brittle. Treatment can be detected by disruption to inclusions, which often have a lower melting point than the host, appearing as 'melted snowballs' or with radiating stress fractures. Some gems require analysis by a gemmological lab to identify heating, but for many, such as aquamarine, morganite, tanzanite, tourmaline and topaz, it is undetectable.

Heating is undertaken on a stove, in a fire, oven or furnace. Factors that affect results include how fast heating and cooling occurred, maximum temperature reached, total length of time, pressure conditions, and if the atmosphere was oxygen-rich (oxidizing) or oxygen-poor (reducing). Some gemstones, such as tanzanite, are heated to a few hundred degrees Celsius. Others undergo much higher temperatures, including corundum which is heated to 1,800°C (3,272°F). Rapidly reducing temperature during treatment can purposely induce fractures in quench crackled quartz, and the disc-shaped stress fractures in amber called sun spangles. Certain diamonds, particularly synthetic chemical vapour deposition stones, are treated by high pressure and high temperature (HPHT) to decolourize or alter the colour. HPHT treatment is becoming more commonly used on sapphire. Heating ruby in a flux such as borax causes the inner surfaces of thin fissures to melt and recrystallize, healing shut. This is frequently used on heavily fractured ruby

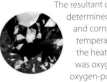

▼ Heat-treated zircons, altered from their original reddish-brown to blue, golden or colourless. The resultant colour is determined by their origin and composition, the temperature used, and if the heating environment was oxygen-rich or oxygen-poor.

from Mong Hsu, Myanmar, and requires specific disclosure in addition to heat treatment.

Diffusion

Diffusion is a process whereby a chemical element is dispersed into a gem's crystal structure during heating. This alters the chemistry of the gemstone and is permanent and stable, so requires specific disclosure.

Diffusion is used for corundum (Al_2O_3). The gem is heated to very high temperatures (1,700–1,900°C or 3,092–3,452°F) in alumina powder containing the diffusion element. The treatment penetrates to different depths depending on the element used, and its relative atomic size to the gem's elements. Titanium creates deep blues, but as the titanium atom is larger than aluminium and oxygen, it cannot penetrate far, known as surface diffusion. This creates a thin blue layer at the surface less than 0.5 mm (0.02 in) deep. The treatment is applied to finished gemstones, and is detected by immersion as the colour follows the facets rather than crystal growth. Recutting or repolishing removes this thin rind, losing the colour. Titanium diffusion can also improve asterism by increasing rutile silk. Nickel, iron and chromium are also used for surface diffusion. The smaller element beryllium penetrates much deeper, and this is called lattice diffusion. It is diffused into pink sapphires to create a yellow outer zone, giving the desirable appearance of pinkish orange padparadscha. Beryllium diffusion of ruby can improve the red colour. The identification of beryllium treatment requires advanced analytical testing, such as LA-ICP-MS, as some analytical techniques do not detect lighter elements like this one.

Irradiation

Exposing a gemstone to radiation can alter its colour. Some gems undergo this process naturally underground by proximity to radioactive minerals, for example smoky quartz. Artificial irradiation replicates this process to produce the same results or creates an unnatural colour. Irradiation takes place in a reactor, using electron, neutron

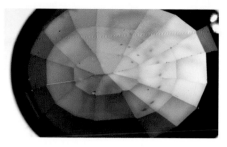

▲ Lattice diffusion of titanium during heat treatment of sapphire is detected by blue colour concentrations following facet edges, seen when immersed in diiodomethane.

or gamma sources. It is often followed by heat treatment, called annealing. This further modifies the colour by removing unwanted or unstable hues, producing a pure colour. One of the best examples is blue topaz, so commonly produced by treatment it is assumed all medium to dark blue topaz on the market is irradiated. Irradiation is also used to treat tourmaline, aquamarine, diamond, pearl and varieties of quartz.

The colour of some irradiated gemstones is unstable, fading with exposure to light. In others, the colour is stable, but heat should be avoided which might revert the colour. This treatment is often undetectable or requires laboratory analysis. Colours rarely seen in nature should be considered suspicious. The sale of irradiated gems is highly regulated to ensure the gemstones do not contain any residual radiation.

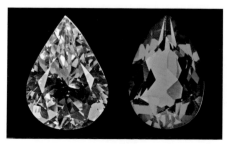

▲ Untreated colourless topaz (left) and topaz treated by gamma irradiation followed by heat to ~200°C to produce the blue colour (right).

Imitations and synthetics

It is only in the last few centuries that advances in science and the understanding of chemistry, composition, and geological processes have allowed for the synthesis of gemstones. Today the use of imitations and synthetics is commonplace, greatly widening the availability of many gems, and at affordable prices. In gemmology, specific terms describe these impersonators. An imitation is a gem of one material (either natural or human-made) that imitates the appearance of another. An artificial gem is one made by humans. A synthetic is an artificial mineral with the same chemical composition and crystal structure as its natural counterpart, thus having the same chemical, physical and optical properties. Because of the production of effective imitations and inexpensive high-quality synthetics, and the difficulty distinguishing them from their more valuable natural equivalents, there are strict requirements for disclosure when marketing and selling. Terms such as 'real' should be avoided – both synthetic and natural emeralds are 'real' emeralds, they only differ in that one is made by humans, and the other by nature.

IMITATIONS

Any gem can be imitated, usually by a more plentiful and/or cheaper substitute of similar appearance. Gem minerals are used as imitations, for instance colourless quartz, topaz, sapphire and zircon have long been used to mimic diamond. Organic materials are similarly utilized, particularly to imitate gems whose trade is banned or restricted. Horn simulates tortoiseshell, coral, jet, and along with vegetable ivory and bone, is carved to appear as ivory. Amber is imitated by the younger resin copal. Natural gems may be treated to adopt the appearance of others. Dyeing is prevalent – calcite to imitate jade, howlite to mimic turquoise and quench crackled quartz to simulate ruby or emerald.

Of the artificial imitations, glass (or paste) is one of the most common and has been used for millennia. Glass is manufactured cheaply in a range of colours and optical effects, imitating any gem from ruby to jade to jet. It provides the heft and vitreous lustre of many natural gems, but its low toughness and hardness (~5.5) means it does not wear as well as most common and commerical gems. Glass is amorphous, lacking directional properties (isotropic). Other tell-tale signs are conchoidal chips, rounded facet edges from moulding, and a lack of natural inclusions, instead containing rounded bubbles, swirls and flow structures. Composition, and therefore properties, range widely, with an RI commonly between 1.50 to 1.70. Many other well-known isotropic gems do not have RIs in this range, helping to identify it.

The addition of lead from the 1600s improved the appearance of glass, increasing RI, brilliance and dispersion, albeit reducing hardness. Additional ingredients give colour, such as blue from cobalt, red from selenium or gold, and a colour change effect from rare earth elements is used to imitate alexandrite and diaspore. Copper inclusions create glittering aventurine glass, and opalescent glass is created with microscopic particles that scatter the light. Fibre optic glass produces a cat's eye effect, while a coating on a gem's back creates a star effect.

Plastic (resin, polymer) is commonly used as an imitation. Similar to glass, it is amorphous, and often has moulding marks and rounded facet edges, however it is warmer to the touch, and

◀ Gilson opal seen under magnification showing distinct colour regions with a regular polygonal pattern described as 'lizard skin'.

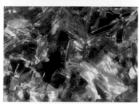

◀ Natural opal also seen under magnification for comparison, showing random variable colour flashes.

lighter in heft. Plastic is used to imitate many gems, including tortoiseshell, coral, turquoise, all types of faceted gems, and beads with an iridescent coating can resemble pearl. Plastic is an effective simulant of amber, encapsulating inclusions of modern plants or animals.

Synthetic gems are used as imitations. Spinel (the great imposter) and corundum are easy and inexpensive to synthesize in precise colours, producing bright hardwearing simulants of aquamarine, topaz, tourmaline, peridot and kunzite to name but a few. An adularescent synthetic spinel mimics moonstone, and a grainy blue synthetic spinel, coloured by cobalt with thin inclusions of gold, imitates lapis lazuli. Cubic zirconia and moissanite, which only occur in nature as small grains, are synthesized to imitate diamond.

In the 1970s, French chemist Pierre Gilson created several convincing imitations of turquoise, opal, lapis lazuli and coral. These artificial products are close to, but not an exact counterpart of, the natural gems. Turquoise, lapis lazuli and coral are made by sintering, that is heating powdered ingredients to nearly melting point and compacting to fuse the particles together. Gilson opal, like natural opal, is made of silicon spheres settled into a regular array, with a convincing play of colour.

COMPOSITE STONES

Composite gemstones are assembled, combining two or more materials, to appear as a single gem. A doublet consists of two layers, while a triplet has three. Composites are created to imitate, such as a garnet-topped doublet (GTD), or increase durability by capping fragile gems. A GTD consists of a thin slice of natural garnet (usually red) fused onto coloured glass, and faceted. This creates a gem the colour of the glass, with the lustre, hardness and inclusions of a natural gem when viewed face up. GTD are distinguished by the garnet/glass join and difference in lustre when viewed in reflected light, although jewellery settings may hide this. The RI will also indicate garnet instead of the expected gem. GTD imitate many gems including emeralds, rubies and sapphires, although today are largely superseded by synthetics.

Other doublets include tanzanite or emerald slices on glass. Natural sapphire or ruby slices topping a synthetic corundum base is a convincing imitation with the same colour, inclusions and RI as a natural stone. Flattened bubbles along the interface reveal the composite. Soudé gemstones are composites using rock crystal, pale beryl or synthetic spinel either side of a coloured layer or cement, usually in green to imitate emerald.

Triplets include glass sandwiching particoloured plastic to imitate tourmaline. Opal triplets consist of a very thin opal slice attached to ironstone to imitate

▶ A soudé emerald, composed of a colourless crown and pavilion cemented together by a green layer to appear green face up.

boulder opal, or black glass, plastic or onyx to create black opal. These are capped by curved rock crystal or glass to protect and magnify the play of colour.

▶ A complete synthetic ruby boule grown by the flame-fusion method, weighing 415 ct and ~9 cm (3in) length.

SYNTHETIC GEMS

Many gems, particularly diamond, emerald, ruby and sapphire, are synthesized. There are several techniques to grow crystals. Each requires the ingredients to provide the necessary chemical elements, appropriate formation conditions often recreating the natural process, and time to grow. Most synthetic gem materials, such as diamond, corundum, fluorite and quartz, are produced for use in industry, ranging from abrasives, electronics, watches, ceramics and optics, with only a minor portion used as gemstones.

▼ Curved striae in a flame-fusion synthetic ruby, with torpedo-shaped gas bubbles perpendicular to the growth lines. FOV 3.59 mm.

While imitations have been used for thousands of years, the synthesis of gems was only achieved in the 1800s, and not at commercial levels until the 1900s. Ruby was the first gem to be synthesized in transparent crystals of facetable size, targeted for both its high value and simple chemistry (Al_2O_3). Following several attempts in the 1800s, the unveiling in 1902 of Auguste Verneuil's flame-fusion method led to its commercial production. Synthetic sapphire and spinel soon followed, and this method is still widely used. It was not until the 1940s that other gems grown by different methods, such as synthetic emerald, joined the market. As synthetic gems have the same properties as their natural counterparts, inclusions (particularly the lack of natural inclusions) are key to their identification.

Flame-fusion (Verneuil) method

The flame-fusion method drops a powdered feed through a high temperature oxy-hydrogen flame. Melted droplets fall onto a rotating pedestal and recrystallize layer upon layer to form a cylindrical crystal called a boule. This inexpensive method grows at ~1 cm (0.39 in) per hour, completing a boule in just a few hours. Most synthetic corundum is made by

this method, using a feed of alumina powder mixed with small quantities of colouring agents (dopants) such as chromium for ruby, and iron and titanium for blue sapphire. Titanium dioxide (rutile) is added to create a star stone, and reheating the completed boule induces the formation of orientated rutile silk. Spinel, similarly simple in composition ($MgAl_2O_4$), lends well to this method and is produced in almost every colour. The process for synthetic spinel requires marginally more aluminium than natural spinel, giving it a slightly higher RI and internal strain seen as distinctive 'tabby' stripes of anomalous extinction on the polariscope.

Flame-fusion synthetics are suspiciously clear, detected by their curved (never straight) growth lines. These are seen under magnification as colour zones or striae, although can be challenging to detect, especially in pale colours. They may also contain gas bubbles, either dispersed along the curved growth zones or elongated at 90 degrees. Fingerprints/feathers may be induced to give the appearance of a natural gem.

Crystal-pulling (Czochralski) method

The crystal-pulling method, developed by Jan Czochralski around 1917, similarly grows crystals from a melt. The ingredients (including colourants) are melted in an iridium crucible, into which an oriented seed crystal is dipped then slowly rotated and withdrawn, pulling a crystal from the melt up to 25 mm (0.98 in) per hour. This creates a large, high-quality crystal almost free of inclusions, mainly used for industrial applications such as ruby for lasers. Rarely it contains elongated bubbles or faint curved growth lines. Gems created this way include corundum, chrysoberyl, alexandrite, the artificial garnets yttrium aluminium garnet (YAG) and gadolinium gallium garnet (GGG), and a strongly pleochroic synthetic forsterite, coloured blue by cobalt, which effectively imitates tanzanite.

Flux melt method

The flux melt (or flux growth) method uses a flux, a solid material such as a metal alloy or oxide that when melted dissolves the ingredients and allows crystals to grow at lower temperatures than normal. The molten flux and ingredients are held in a platinum, iridium or graphite crucible,

and as the solution slowly cools, crystals nucleate spontaneously, or on a natural or synthetic seed crystal. Further ingredients may be added to the crucible as required. This method grows distinctive crystals with faces, therefore flux melt gems may show angular zoning. Inclusions of undissolved flux occur as breadcrumb-like particles, paint splashes or feathers/twisted veils resembling smoke or, confusingly, natural liquid fingerprints. Other inclusions are seed crystals and metallic platelets from the crucible. There are no liquid inclusions, and no natural solid inclusions.

The flux melt process is carefully controlled, taking up to a year to grow crystals. While time-consuming, and expensive, the resultant gems are of fine quality. This process can produce emeralds, rubies, spinel (including red, which is difficult to produce by flame-fusion), alexandrite, and YAG. Several companies create flux melt gems, and they are commonly named after the manufacturer, for example Chatham-created emerald and ruby.

Hydrothermal method

The hydrothermal method recreates the natural geological conditions to grow crystals from a solution. Water, rather than a flux, dissolves the ingredients under heat and pressure in a sealed container called an autoclave. As the solution slowly cools, crystals form on a suspended seed crystal. This method is slow and expensive, but faster than flux melt. As the autoclave is sealed, no further ingredients can be added. This process is mostly used to produce emerald, ruby and quartz, particularly amethyst. Hydrothermal alexandrite, ruby, sapphire, morganite and red beryl are also seen on the market. Hydrothermal synthetics

▲ A flux melt synthetic ruby with crystal faces, quite different in shape to natural ruby crystals.

▶ A colourless synthetic quartz crystal grown by the hydrothermal method.

may be eye clean, with chevron zoning, and a roiled internal appearance. They can be difficult to distinguish from natural gemstones.

High pressure high temperature (HPHT) method

The HPHT method simulates diamond's natural formation conditions, creating high pressure through a split sphere multi-anvil press or a high pressure belt apparatus. A central growth chamber contains a flux in which industrial diamond powder or graphite is dissolved, with or without colouring agents such as nitrogen (yellow) or boron (blue). The diamond grows from the dissolved carbon onto a seed crystal of natural or synthetic diamond, at temperatures lower than required in nature. The first HPHT diamonds were successfully grown by a reproducible method in the 1950s. Most HPHT diamonds fluoresce, and are distinguished from natural diamond by a stronger reaction under shortwave UV light than longwave UV light. They additionally lack internal strain when viewed on the polariscope. The HPHT process is also used to remove brownish colours from both natural and chemical vapour deposition diamond, and recently to treat sapphire.

Chemical vapour deposition (CVD) method

CVD diamond is grown in a chamber containing methane (CH_4) and hydrogen (H_2) vapours. Through a chemical reaction, the carbon atoms precipitate out of methane and are deposited in layers on a seed plate or substrate to form a single crystal. The process occurs under high temperature, low pressure conditions. CVD diamond was first achieved in the 1950s, but not commercially produced until the late 1990s. It may be identified using the polariscope by an anomalous columnar pattern from internal strain, but often requires laboratory analysis.

Analysis

The analysis of a gemstone determines its composition, if it is natural or synthetic, and if it has undergone treatment. Analysis can also indicate the geological setting of formation and sometimes the geographical location. However it is not a straightforward process. Gem materials are hugely diverse – minerals, rocks, organics, human-made. In turn, each material can vary, ranging in physical and optical properties. New gem minerals, and deposits, continue to be found.

Analysis is further complicated by undetectable treatments, new treatments appearing undisclosed on the market, and synthetics with the same properties as their natural counterparts. Testing is usually limited to non-destructive methods, and only part of a gemstone may be accessible when set into jewellery. Gemmology is a constantly evolving field that must keep pace with these new advances. As such, a gemmologist must turn detective, using multiple methods to build evidence to establish identity.

A gemmologist's greatest tool is observation, by eye, or under magnification first by hand lens then microscope. Different lighting setups, provided on gemmological microscopes, are essential for viewing different types of gemstones, or to highlight the inclusions within.

◀ A gemstone being observed on a gemmological microscope, illuminated from beneath.

The observation of external features provides many clues as to a gem's identity. Gems can have characteristic lustre, and very bright lustre may indicate a high refractive index (RI). Wear, such as scratches, suggests low hardness, and gems of poor toughness are more likely to show damage. Fracture may be characteristic, such as conchoidal chips in quartz or glass. Rounded facet edges may indicate moulded glass, or a relatively soft gem, while sharp edges with flat facets implies a hard gem such as diamond, which takes a high polish. Treatments may also be observed – wear on coatings, colour concentrations of dye in cracks, different surface lustre of fracture fillings, and join lines on composite gemstones.

▲ Abraded facet edges with relatively scratch-free facet faces, seen under magnification, reveals the identity of this zircon.

Internal features, particularly inclusions, give vital clues to identity, treatments and origin. Inclusions are directly related to the process and environment of formation, in both natural and human-made gems. Curved zoning in corundum or spinel distinguishes a flame fusion synthetic gem, while straight colour zoning may indicate natural. Calcite and pyrite in an emerald suggest Colombian origin, while mica indicates a schist-hosted deposit, such as Zambia or Brazil. Some inclusions are considered diagnostic, like 'rain' in aquamarine, or 'horsetails' in demantoid garnet. Stress fractures around inclusions imply heating.

The specific gravity of unset gemstones can be measured using a hydrostatic balance. This is based on Archimedes Principle, and measures the weight of the gem in air, and submerged in water, to determine the relative density. While fiddly to measure, specific gravity is useful to distinguish similar gems such as nephrite and jadeite jade.

The testing of a gem's optical properties, to determine how it interacts with light, gives clues to its internal structure, composition and optical density. Testing by handheld gemmological equipment is non-destructive and relatively simple with practice.

The polariscope uses plane polarized light to analyze transparent and translucent gemstones, distinguishing isotropic (singly refractive) from anisotropic (doubly refractive) materials. It consists of two polarizing filters, which sit over a light source. The filters are set one above the other in crossed polarization, that is rotated at 90 degrees. The plane polarized light transmitted through the bottom filter is not able to transmit through the top filter. The gemstone is placed between the two, acting as its own filter for the light transmitted through the bottom filter. When rotated, isotropic gemstones remain dark (called extinction) and anisotropic gems appear dark (extinct) then light every 90 degrees. Anomalous extinction effects occur due to internal strain in isotropic gems, and may be diagnostic. Natural diamonds are strained from their turbulent formation, while synthetic ones are not.

◀ The polariscope, with two polarizing filters, one above the other, set at 90 degrees to each other, with a gemstone placed between.

▼ The refractometer with a gem placed face down on the small glass table. The yellow light source behind shines into the refractometer, and the scale is viewed through the eye piece at the front.

For doubly refractive minerals, each ray has its own shadow edge for which its RI can be determined (see double refraction, p21). The maximum difference between the two RIs is the birefringence. More information on the optical nature is gathered by rotating the gemstone. Observing if, and how, one or both shadow edges move, indicates if the gem is uniaxial or biaxial, with a positive or negative optical sign. This can distinguish between gemstones with overlapping RI values such as quartz and scapolite. Natural emerald and spinel may be distinguished from synthetic by slightly differing RI and birefringence values.

The dichroscope is a simple but effective tool helping to identify gems by their typical pleochroic colours. It tests transparent, coloured gemstones by transmitted light. The London dichroscope comprises two polarizing filters set side by side at 90 degrees. When viewing a doubly refractive gem, each filter blocks out one of the two transmitted rays, allowing them to be seen individually, showing each one's pleochroic colour side by side. The detection of pleochroism also rules out singly refractive materials, for example distinguishing ruby from garnet, red spinel and glass.

The Gem-A Chelsea Colour Filter only allows the transmission of light in deep red and yellow-green wavelengths. This means only gemstones that

Interestingly, synthetic spinel shows anomalous shadowy stripes from strain while natural spinel remains dark.

The RI of a gemstone is determined using a refractometer. As each wavelength of visible light has a different RI (see dispersion, p21), the RI of a gem is measured using one wavelength, 589 nm, known as sodium yellow. This allows comparison between different gems. The refractometer uses the concept of total internal reflection which is determined by a gemstone's critical angle. The gemstone is placed on the refractometer's small glass table (which has a higher RI), using a drop of RI liquid at the interface to ensure optical contact. The yellow light shines internally through the glass and, dependent on the angle of incidence, is refracted into the gem or totally internally reflected, illuminating a scale. The division of refracted and reflected light is determined by the gem's critical angle and is seen on the scale as the edge between bright reflected light, and dark shadow. The value indicated by the shadow edge is the gem's RI. The contact liquid normally has an RI of 1.79 or 1.81 and this constrains the upper limit of testing. Gems of higher RI, such as spessartine garnet and diamond, are therefore beyond the limits of the standard refractometer.

▶ The Gem-A Chelsea Colour Filter distinguishes between these two pale blue gems. The aquamarine (left) appears greenish blue, whilst the synthetic spinel (right) appears red. Note, this is a composite photograph to allow the effect to be seen in print.

transmit light in the deep red wavelengths appear reddish through the filter. Green gems containing chromium, such as emerald and demantoid garnet, and blue gems with cobalt, such as blue spinel and glass, also transmit red wavelengths, thus appearing red. This simple tool can quickly separate similarly coloured gemstones, such as aquamarine (appearing pale greenish blue), blue topaz (appearing pale orange), blue glass and blue synthetic spinel (appearing red). It can also indicate treatments such as chromium-containing green dye in jadeite jade.

A spectroscope is a tool that spreads visible light into the full spectrum. This reveals which wavelengths are transmitted or absorbed by a gem. The transmitted light enters the spectroscope through a slit and is dispersed by a diffraction grating or prism. The absorbed wavelengths are removed from the spectrum, creating dark lines or bands. This pattern is known as an absorption spectrum. Research has determined which wavelengths are absorbed by each chemical element, and some chromophores, such as iron, give a distinctive pattern. While a handheld spectroscope is not sensitive enough to detect exact wavelengths, the pattern is enough to indicate the element causing the colour. Many gems have a characteristic spectrum that can identify them.

Fluorescence, seen under longwave and shortwave UV light, can be characteristic, and is a useful tool to identify a gemstone, indicate treatment, and distinguish synthetic origin.

At times, a gemstone requires more sophisticated testing, beyond the capabilities of standard gemmological equipment. Major advances over the past 40 years in treatments and synthetics, coupled with increasing demand for geographic origin determination, necessitates the use of complex analytical methods. These are normally undertaken by gemmological labs. Techniques are developing continuously, with equipment modified for non-destructive micro- to nanoscale analysis. These methods allow the detection of treatments on an atomic scale, identification of high-quality synthetics, detailed chemical and structural assessment, and accurate analysis of trace elements down to parts per billion. The trace element chemistry of a gem reflects the geological environment of formation, creating a geochemical fingerprint that can be used to determine geographical origin.

X-rays, used to view bones inside the body, are also used on opaque gemstones such as pearl to reveal their internal characteristics. X-ray analysis can distinguish natural versus cultured, and the presence or absence of bead nuclei. Testing transparent gemstones can identify fracture fillings, as the lead glass used in diamond and ruby is more opaque to X-rays. A scanning electron microscope (SEM) uses a beam of electrons to gain information from a sample at nanoscale – used to discover that the structure of opal was an array of spheres.

Sophisticated spectrometers can analyze wavelength absorption from the ultraviolet through visible light to near infrared (UV-Vis-NIR). This produces an absorption spectrum detecting the exact wavelength, and the intensity, of absorption. As the absorption directly represents the elements and defects in the gemstone that cause colour, it can identify the gem material, cause of colour and indicate the geological setting and geographical origin. This process is commonly used as a first step to separate basaltic (e.g. Australian) and metamorphic (e.g. Sri Lankan) blue sapphires, and emeralds of hydrothermal/ metamorphic (e.g. Colombian) or schist-hosted/ magmatic (e.g. Zambian) origin.

Raman spectrometers analyze the structure of a material by hitting it with a laser and measuring the scattered laser light. This produces a Raman spectrum, characteristic of the gem material, which is identified by comparison to known spectra.

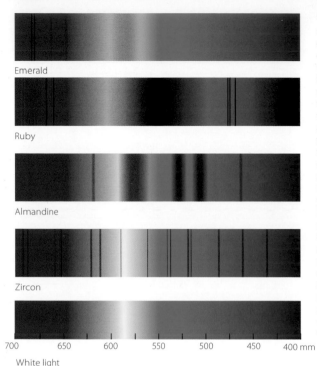

Emerald

Ruby

Almandine

Zircon

700   650   600   550   500   450   400 mm

White light

◀ Characteristic absorption spectrums of different gems, seen with a handheld diffraction grating spectroscope. From top to bottom: natural emerald spectrum (from chromium) with a doublet in the red, two weak narrow bands in orange and weak absorption in the yellow, but transmitting all of the green; ruby spectrum (from chromium) with emission a doublet in the red, two lines in orange, broad absorption of yellow and green, and three lines in the blue; almandine spectrum (from iron) with three characteristic bands in at 576 nm, 527 nm and 505 nm; zircon spectrum (from uranium) with many fine lines and weak bands seen in an untreated stone; white light spectrum (no absorption) with nanometre scale of the wavelengths of visible light.

Raman microscopes can analyze minute particles, such as inclusions, establishing identity even when they are beneath the surface of the gem. Raman is often used in combination with methods which determine chemistry, such as X-ray fluorescence.

Energy dispersive X-ray fluorescence (EDXRF) determines the composition of a gem by measuring the emission of X-rays. The gem material is energized by X-rays, or an electron beam (in a scanning electron microscope), rereleasing the energy as X-rays. This process identifies elements, estimating their relative abundance, however it cannot detect elements lighter than sodium. This non-destructive technique is widely used, for example to distinguish salt and freshwater pearls by their manganese content, and detect silver nitrate treatment.

Following the appearance of beryllium-diffused sapphires in the early 2000s, analytical methods that detect the lighter elements were required. Laser ablation inductively coupled plasma mass spectrometry (LA-ICP-MS) detects elements as light as beryllium in very low concentrations, down to parts per billion. It ablates a tiny spot about half the width of a human hair (so is considered semi-destructive) then uses a mass spectrometer to determine the elements and their quantity. This is one process used to determine the geographic origin of the copper-bearing Paraíba/Paraíba-type tourmalines. These beautiful gemstones can sell for as much as US$150,000 per carat (£108,000) when originating from Paraíba, Brazil, while equivalent gemstones from Africa are valued at a quarter of that or less. As the source has a serious impact on value, accurate analysis of low levels of trace elements is essential to determine origin.

The gemstones

The following two chapters describe the most commonly encountered gems, as well as some of the lesser known. These gems have geological origins and are formed from minerals and rocks. The third and final chapter discusses organic gems, grouped separately due to their biological origins.

It is important to understand the scientific classifications for gems of geological origin:

- A mineral species (or simply mineral) is a natural, inorganic solid with a specific chemical composition and crystal structure. The list of approved mineral species is governed by the Commission on New Minerals, Nomenclature and Classification (CNMNC), part of the International Mineralogical Association (IMA).

- A mineral variety shows a distinct variation in appearance, crystal shape or optical phenomena. The variety name may also indicate the source, or have another meaning. The variety has the same internal crystal structure, with the same or slightly varied composition but not varied enough to be defined as a new mineral. For example, ruby is the red variety of the mineral corundum, and Paraíba tourmaline is the copper-bearing variety from Paraíba state in Brazil. Variety names are not governed, and range widely. Some varieties, such as moonstone, occur in more than one mineral species.

- A mineral group comprises closely related mineral species, usually with the same crystal structure but different compositions. As it can sometimes be challenging to distinguish individual mineral species, some gems, such as tourmalines, are known by their group name.

- A mineral solid solution series is a continuous variation in composition (and therefore properties) from one mineral species (called an end member) to another, usually due to one element being increasingly substituted by a different element until the first mineral becomes the second. The garnet group forms solid solution series, and as garnet gems are often compositional blends, it can be difficult to determine the exact mineral species.

The categorization of gems is complicated. The name by which a gem is known may be the scientific name (mineral species or group), a colour variety, or even a trade name. The sequence for discussing gems is also not standardized – they may be ordered alphabetically, or by importance and popularity.

In this book, I have chosen to order the well-known and lesser-known gems alphabetically by mineral species, so they are easy to find. Within each species, varieties are discussed, and where the varieties are well known and/or have slightly varying properties they are given their own sections. These are ordered by their colour varieties or optical phenomena, accounting for mineralogical relationships such as solid solution series. Quartz is divided into its macrocrystalline and cryptocrystalline (chalcedony) varieties.

The exceptions are the mineral species of the feldspar, garnet, and tourmaline groups, plus jade (a gemmological term for two similar gems). These are discussed by group rather than species, due to their close similarities, difficulty in distinguishing between individual species, lack of pure compositions, and overlapping varietal names. Within these groups, the important gem species or varieties are given their own sections. The properties listed include the most commonly observed ranges for each gem.

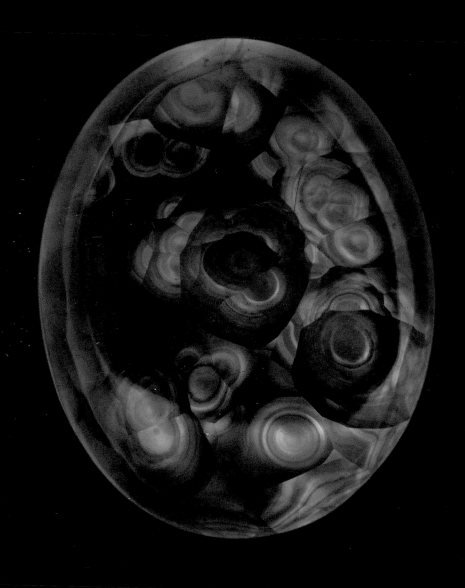

Well-known gems

From diamond and ruby to tourmaline and zircon, this chapter discusses the most commonly encountered gemstones and their well-known varieties. These geological gems are favourites for adornment, each having a unique combination of properties that creates a sparkling allure.

Andalusite

Composition Al_2SiO_5 aluminium silicate
Crystal system orthorhombic **Hardness** 7–7.5 **Cleavage** good **Fracture** uneven, subconchoidal **Lustre** vitreous to greasy
SG 3.05–3.21 **RI** 1.627–1.651 (viridine 1.690)
Birefringence 0.007–0.014 (viridine 0.029)
Optical nature biaxial - **Dispersion** 0.016

◀ A waterworn crystal of andalusite from Brazil.

Andalusite is an attractive green to greenish brown gem, occasionally red to pink. It was named in 1798 after the Spanish region Andalusia, where the specimens were mistakenly thought to originate. They actually came from El Cardoso de la Sierra in Castile-La Mancha, a region to the north. Intriguingly the well-known variety chiastolite was already known, described also from Spain 44 years earlier.

Occasionally found in pegmatites, andalusite occurs most commonly in metamorphic rocks. It is a polymorph of Al_2SiO_5, together with kyanite and sillimanite, meaning they share the same composition but have different crystal structure. Each polymorph grows under separate pressure and temperature conditions, with

andalusite forming in low pressure and low to high temperature. As they rarely occur together in the same rock, these polymorphs are used as indicators of the metamorphic conditions undergone by the host rock. Gem-quality andalusite is rare and most is found as waterworn crystals or cleaved sections within gem gravel placer deposits. The main sources are Brazil, Sri Lanka and Myanmar, with minor sources including Madagascar and China. South Africa produces andalusite for industry, used in furnaces due to its ability to withstand high temperatures.

◀ Beautiful iridescence from the layering within this fire agate.

Andalusite has orthorhombic crystal symmetry. It forms equant or elongate prisms with a squarish cross section, or can be fibrous or massive. It has cleavage in two directions at almost 90 degrees, which reduces its toughness, and this combined with its rarity means It is normally a collector's gemstone. Inclusions may be brown mica flakes, clear crystals, bladed or tubular parallel inclusions, or incipient cleavage.

This gem's beauty comes from its strong pleochroism, best seen in green or reddish material. Its trichroic colours of reddish brown/green/yellowish green can be seen easily in both crystals and cut gemstones when viewed from different directions. Gem andalusite is typically transparent to translucent with a vitreous lustre, and most is faceted. The rough is cut to show two or more pleochroic colours when the gemstone is viewed face up. The attractive flashes of red and green, or yellow and green, are so desirable that stones are cut to achieve this to the detriment of carat weight. Elongated rectangular or oval shapes are common, showing one colour in the middle and another towards the ends, while rounded cuts appear multicoloured. Most gemstones are under 20 ct although larger stones are known.

There are two gem varieties of andalusite. A beautiful deep green manganese-bearing variety occurs from Brazil and is known as viridine. The other is chiastolite or cross stone, a very different, mostly opaque variety with a distinctive cross-shaped internal pattern. These crystals are cut into

▲ A chiastolite crystal section with a polished face from Bimbowrie, South Australia.

polished slices or cabochons. They have a pale white, grey, yellowish or pinkish body colour and a black cross caused by carbonaceous inclusions. Chiastolite is well known from Spain, Australia, and the USA, with other sources including Russia.

Andalusite is generally not treated; rarely it may be heated to improve colour, or fracture filled to improve the clarity. Cleaning is best with warm soapy water. Ultrasonic and steam cleaners should be avoided due to the risk of cleaving. Andalusite may appear similar to tourmaline, especially when cut to show the pleochroism, but can be distinguished on the refractometer by its different birefringence. It may also be confused with the similar colours of peridot and chrysoberyl, but these lack the strong pleochroism.

▶ A 23.96 ct step cut andalusite from Minas Gerais, Brazil. It shows strong pleochroism, with a different colour seen in each direction. Left to right: top view (green), side view (yellowish green), end view (reddish brown).

Apatite

Composition $Ca_5(PO_4)_3(F,OH,Cl)$ calcium phosphate **Crystal system** hexagonal **Hardness** 5 **Cleavage** indistinct **Fracture** conchoidal **Lustre** vitreous **SG** 3.10–3.35 **RI** 1.628–1.662 **Birefringence** 0.002–0.010 **Optical nature** uniaxial - **Dispersion** 0.013

Popular with mineral collectors, apatite occurs in a surprising variety of colours, and makes attractive gemstones. Its low durability however limits its use in jewellery. The name apatite comes from the Greek word meaning to deceive, as it was often mistaken for other minerals, such as beryl, due to its colour range and crystal shape.

Perhaps most interesting is the role apatite plays in our lives – it is one of the few minerals found in biological systems. Apatite is the mineral component of our bones and teeth enamel. In fact, fluoride is added to drinking water to prevent tooth decay due to the way it reacts with the apatite in teeth enamel, making it more resistant to acid attack. The main use of apatite in industry is in the manufacture of fertilizer.

Apatite is a calcium phosphate, and mineralogically speaking is a group of three mineral species distinguished by their concentrations of the fluoride (fluorapatite), hydroxyl (hydroxylapatite), and chloride (chlorapatite) ions. Fluorapatite is the most prevalent, however the generic name apatite is normally used for all in the gem industry.

Part of the hexagonal crystal system, apatite forms six-sided elongate, stubby or tabular prismatic crystals. These typically have flat or pyramidal terminations, and striations down the sides. Crystal specimens are sought by mineral collectors for their beauty.

Apatite is a common mineral found in many types of rock – igneous, metamorphosed skarns or marbles, and phosphatic sedimentary rocks to name a few. Gem-quality crystals are mostly associated with pegmatites or hydrothermal veins. They may also be found in alluvial placer deposits.

Significant sources include Brazil, which produces blue, yellow and green stones, and Mexico for yellow stones. The Panasqueira Mine in Portugal is noted for its green 'asparagus stone' and purple apatite. Other sources include the Kola Peninsula in Russia, Myanmar, Canada, and several locations in the USA, including Maine famed for its purple apatite. Madagascar produces neon blue-greens that are highly desirable with a similar colour to Paraíba tourmaline. A blue-green variety known as moroxite was first found in Arendal, Norway. Gemstones showing a chatoyant cat's eye effect come from Myanmar, Brazil, Sri Lanka and Tanzania.

◀ Hexagonal apatite crystal, 5 cm (2 in) long, with pyramidal termination from Durango, Mexico.

▶ A desirable purple 1.08 ct step cut apatite from Auburn, Maine, USA, with numerous tiny inclusions.

▶ Deep blue step cut 1.24 ct apatite from Brazil.

▼ Oval step cut green 1.99 ct apatite from East Africa.

▶ A 16.56 ct apatite cabochon from Myanmar, with parallel needle-like inclusions creating a diffuse cat's eye effect.

Apatite forms in a wide range of appealing colours – blue, green, yellow, purple or pink – and colourless. Transparent material is normally faceted. It is rarely eye clean and may have inclusions of hollow tubes and rehealed fractures. Material with many parallel needle-like inclusions is chatoyant and may be cut en cabochon to produce a cat's eye effect, commonly found in yellow or green colours. Some apatite fluoresces, at times caused by impurities of rare earth elements.

Colour is the most desirable aspect of apatite, and stones are cut to make best use of it. The elongate to tabular shape of the rough will also determine the shape of the gemstone.

Round, oval or baguette shapes are typical. The majority of faceted gemstones are small. More opaque material may be polished into beads, or decorative items.

Apatite is a soft mineral, easy to scratch with a hardness of 5, and is also brittle. Due to this low durability it is usually cut as a collector's gemstone. It may be worn in jewellery but requires protective settings, best suited to earrings and pendants. Apatite is very sensitive to heat and susceptible to chemical attack, thus is not suitable for everyday wear and should be treated with care. The colour of apatite is commonly improved by heat treatments, particularly in blue stones.

Azurite

Composition $Cu_3(CO_3)_2(OH)_2$ copper carbonate
Crystal system monoclinic **Hardness**
3.5–4 **Cleavage** perfect **Fracture** conchoidal
Lustre vitreous **SG** 3.70–3.90 **RI** 1.720–1.850
Birefringence 0.108–0.110 **Optical nature**
biaxial + **Dispersion** none

▼ Tabular azurite crystals intergrown in a rosette, from Boomerang Mine, Queensland, Australia.

The splendid azure blue colour gave azurite its name in 1824. It has been known since ancient times under different guises, and is mentioned in Roman philosopher Pliny the Elder's *Historia Naturalis*, completed in 77 AD. Due to the rich colour, azurite was used as a blue pigment, however the toxicity of the dust and tendency to turn green over time led to other blue pigments being favoured, particularly Prussian blue.

Azurite is a copper carbonate mineral, and is a minor ore of copper. As a secondary mineral, it is found on the outskirts of copper deposits formed by the weathering and alteration of the copper ores. Azurite has a monoclinic crystal structure, occurring in a variety of prismatic and tabular crystal forms, often in complex or radiating groups. Crystals are normally opaque with vitreous to dull lustre. On occasion they are translucent to transparent with electric blue colours seen through thin crystal edges. Azurite also forms as wavy layers, botryoidal clusters or stalactitic forms, and powdery massive material.

This mineral was known through the 1800s as chessylite after the magnificent crystals found in the copper deposits of Chessy-les-Mines in France. Exceptional crystal specimens also come from Tsumeb in Namibia, Millpillas in Mexico, Bisbee in Arizona, USA, Touissit in Morocco, and Peru, and are highly sought by collectors. Other sources include Australia, Germany, Greece and Russia.

Copper is the cause of the pale, bright or deep blue colour for which azurite is famed. Bright blue wavy layering is revealed on polished surfaces and it is used for decorative stones and ornamental objects. To highlight the beauty of the opaque banding, and also due to the very low hardness of 3.5–4, gemstones are typically cut as cabochons, beads or flat slices. Gem-quality transparent crystals may be faceted but are very rare.

Azurite is commonly associated with malachite, a green copper carbonate of similar composition. They are often intergrown with appealing interwoven bright blues and greens known as azurmalachite. This opaque material is fashioned into cabochons or decorative objects. As malachite is more stable than it, azurite crystals frequently alter to malachite forming a pseudomorph retaining the shape of the original crystal. Azurite should be treated with care as it will easily scratch. It is sensitive to heat and may also fade with exposure to bright sunlight. As a carbonate mineral, it is susceptible to acids, so avoid contact with household cleaners, perfumes and hairspray. Azurite may be coated with a colourless layer to improve its hardness and lustre. A reconstructed azurmalachite of compressed azurite and malachite is known, which is stabilized by impregnating with resin to create an attractive, more durable material.

Beryl and its varieties

Beryl is best known for its emerald and aquamarine gem varieties, but also includes morganite, heliodor, goshenite and the rare red beryl. They are prized for their colours, durability, and transparency, creating beautiful gemstones. Emerald is considered one of the most important coloured stones.

Beryl species

Composition $Be_3Al_2Si_6O_{18}$ beryllium aluminium silicate **Crystal system** hexagonal **Hardness** 7.5–8 **Cleavage** indistinct **Fracture** conchoidal **Lustre** vitreous **SG** 2.60–2.90 **RI** 1.560–1.602 **Birefringence** 0.003–0.010 **Optical nature** uniaxial - **Dispersion** 0.014

Beryl is a beryllium aluminium silicate, and as beryllium is a rare element, found in the upper continental crust, this mineral is relatively rare. The gem varieties are generally distinguished by their attractive colours, caused by traces of different chemical impurities. Goshenite, aquamarine, heliodor and morganite are often found in the same deposits and have similar features. They primarily form through pegmatitic processes associated with large igneous intrusions. Here, rare trace elements like beryllium become concentrated in the late stages of crystallization in quantities sufficient for beryl to form. Cavities in pegmatites provide space for magnificent large crystals to grow. Red beryl occurs in beryllium-rich volcanic rhyolites, and emerald forms in schists and veins, through processes involving hydrothermal fluids.

Beryl has hexagonal crystal symmetry, and all varieties occur in well-formed, six-sided elongate to tabular prisms. Non gem-quality beryl can grow to enormous sizes of several metres, and is the main ore of beryllium. The largest documented crystal was found in Madagascar, measuring 18 m (60 ft) long! Gem-quality crystals, particularly of aquamarine and heliodor, can be large, while the rare red beryl grows only to a few centimetres. As a durable gem mineral, beryl may be weathered and transported from its host, accumulating in secondary placer deposits such as gem gravels.

◀ The colour varieties of beryl (left to right) heliodor (yellow), emerald (green), aquamarine (blue), red beryl (red) and morganite (pink), all showing hexagonal form.

Pure beryl is colourless (goshenite). Impurities of iron creates blue (aquamarine) to yellow or green (heliodor), manganese creates pink (morganite) to red (red beryl), and chromium and/or vanadium imparts intense green (emerald). Inconsistency in impurities causes a variation in the depth of colour, and can even create different beryl varieties in the same crystal. These impurities are also responsible for the slight variations in other properties. All coloured beryl has distinct pleochroism exhibiting two variations of the body colour.

Gem-quality beryl ranges from transparent to translucent. Goshenite, aquamarine, heliodor and morganite gemstones are normally eye clean, while emerald and red beryl are rarely without visible inclusions. Inclusions are characteristically fluid or multiphase reflecting the liquid-rich environment in which the crystal grew. Thin, elongate, hollow or liquid filled tubes parallel to the length of the crystal are typical.

Beryl is faceted in a range of cutting styles to highlight the colour. Aquamarine, heliodor and morganite are often pale, so to maximise colour are deeper cut, often with a larger table facet. The rough is orientated with the strongest pleochroic

▶ Pink hexagonal morganite crystal capped by an aquamarine termination.

colour seen face up in the finished gemstone. Material with more inclusions may be fashioned into cabochons. When parallel growth tubes occur in great enough quantity, beryl may exhibit a chatoyant cat's eye effect, or rarely asterism, when cut en cabochon.

This durable gemstone has a hardness of 7.5–8 and has an indistinct cleavage, although it can be brittle. Unless treated by fracture filling, it is resistant to chemical attack. All beryl varieties, except emerald, can be confidently worn every day and are suitable in all types of jewellery. Gemstones with few inclusions are safe to clean by ultrasonic or steam cleaners, but emerald should be treated with extra care as it is more brittle than other varieties, is usually included, and typically treated by fracture filling.

Goshenite

Composition $Be_3Al_2Si_6O_{18}$ beryllium aluminium silicate **Crystal system** hexagonal
Hardness 7.5–8 **Cleavage** indistinct **Fracture** conchoidal **Lustre** vitreous **SG** 2.60–2.90
RI 1.561–1.602 **Birefringence** 0.004–0.009
Optical nature uniaxial - **Dispersion** 0.014

Goshenite is the colourless to near colourless variety of beryl. It is named after Goshen, Massachusetts, USA, where it was discovered, and forms in elongate prisms to short tabular crystals. It ranges from white opaque to transparent with most faceted gemstones eye clean. Goshenite

▼ Goshenite from Minas Gerais, Brazil, ~7.2 cm (2.8 in) wide.

creates reasonably lively stones with vitreous lustre, however it lacks fire due to the low dispersion. It is not often seen as a gemstone in its own right, but has been used to imitate diamond or emerald using a silver or green foil backing. Goshenite can be easily discerned from diamond by the lower hardness and lack of fire. Those imitating emerald are more difficult to distinguish, as they have almost the same properties, however goshenite lacks emerald's inclusions, and the foil backing may show crinkling or discolouration under magnification.

Gem-quality goshenite is relatively abundant, often found with aquamarine. Sources include Brazil, Pakistan, Canada, Russia, China, Myanmar, Sri Lanka and Namibia. It may be irradiated to create other colours, with the resulting colour dependent on the impurities it contains. However, it is generally untreated. It may be confused with other colourless gemstones, such as quartz variety rock crystal, topaz, sapphire and zircon, but can be distinguished by differing refractive indices and specific gravity.

Emerald

Composition $Be_3Al_2Si_6O_{18}$ beryllium aluminium silicate **Crystal system** hexagonal **Hardness** 7.5–8 **Cleavage** indistinct **Fracture** conchoidal **Lustre** vitreous **SG** 2.64–2.80 **RI** 1.560–1.602 **Birefringence** 0.003–0.010 **Optical nature** uniaxial - **Dispersion** 0.014

Emerald, the vibrant green variety of beryl, has captured the hearts of many for thousands of years. It is so well known for its colour that the term emerald green is applied to intense saturated greens. It is one of the most important gemstones alongside diamond, ruby and sapphire, long given high status as a sacred stone with healing powers. Emerald has been known from Egypt for over 2,000 years, and 'smaragdus,' meaning green stone, in Pliny the Elder's *Historia Naturalis* (completed in 77 AD) is interpreted by many historians to be emerald. The name has evolved indirectly from the Greek 'smaragdos' to the English emerald.

This gemstone was recognized as a variety of beryl, rather than a mineral in its own right, only in the late 1700s. It ranges from bluish green to yellowish green, with the most valued an intense pure green or slightly bluish green. The colour is caused by trace impurities of chromium and/or vanadium, which substitute for aluminium in the crystal structure. Iron imparts bluish or yellowish tints. It is generally accepted that beryl cannot be the emerald variety unless coloured by chromium/vanadium. Lighter, less saturated greens, caused only by iron impurities, are the lower value green beryl discussed here under heliodor. Some gemmologists and dealers, however, define emerald by the depth and saturation of green colour regardless of chemical composition. Emerald is pleochroic bluish green/yellowish green. It may fluoresce weak pink to red due to the chromium, although any iron content will quench the response.

▶ Rich green emerald crystals with calcite on black shale, from Muzo, Colombia.

Emerald typically forms elongate, short to long, six-sided prisms with flat terminations. It mostly occurs as single crystals but may grow in radiating aggregates. Although big emerald crystals are known, few occur in large sizes like the pegmatitic beryl varieties, and gem-quality emerald crystals are typically less than 10 cm (4 in) long.

Different to other beryl varieties, emerald is constrained to the unusual geological environments where beryllium, chromium and vanadium are brought together. Deposits occur in contact zones of pegmatite and quartz veins intruding into mafic/ultramafic rock, in dark metamorphic schist either associated with intruding pegmatite or with regional metamorphism, and tectonically folded and fractured sedimentary black shales. In these different environments, hydrothermal fluids circulate bringing the elements together allowing emerald to crystallize in veins, cavities or schist.

As the geological settings are often tectonically active, this causes the emerald to contain fractures (often rehealed) and many inclusions. Crystals may be found as slightly separated pieces within their host rock. Due to these 'defects', emerald is less durable. It does not survive weathering and transport, so is rarely found in secondary placer deposits.

Egypt was the main source of emerald in ancient times, mined in the Eastern Desert from 330 BC, although some historians think as early as 1900 BC. These deposits are known as Cleopatra's Mines, for emerald was said to be her favourite gem (although this is possibly misinterpreted and was peridot, a different green gemstone). Emerald from Egypt is generally pale and included, and considered to be of lesser quality than that from other locations.

Colombia is the famed emerald source, producing incredible crystals and setting the standard for colour. Here, emerald formed from hydrothermal fluids in tectonically faulted black shales in

▲ A cushion mixed cut 1.82 ct emerald, with good clarity, probably from Colombia

limestones, and is often associated with calcite and pyrite crystals. It was mined for centuries by the indigenous Muzo people and traded with the Mayans, Incas and Aztecs. The Spanish introduced these magnificent emeralds to Europe in the 1500s. Today Colombia is still the biggest producer of emerald, and its gemstones are still considered the highest quality. Muzo, Chivor and Coscuez Mines are the most well known.

Zambia hosts large emerald deposits in the Kafubu area, which produce around 20% of the world's production. The Kagem Mine is the primary source. Emerald was discovered in the area in 1928, but not commercially mined until the late 1960s. The crystals are found in contact zones between pegmatites and veins intruded into schists, and some encased within quartz lenses have produced magnificent mineral specimens. Zambian emerald has a bluish tone in comparison to Colombian emerald (due to a higher iron content), and better clarity. Other significant producers from the African continent include Zimbabwe (known for small, but fine, bright green emeralds from Sandawana), Nigeria, Madagascar, Tanzania, Mozambique and South Africa, which is host to the world's oldest emerald at 2.97 billion years. In recent years Ethiopia has produced fine emerald to rival Colombia's.

◀ Emerald crystal in schist from Malyshevo, Ural Mountains, Russia.

▶ Elongate, bamboo-like, amphibole inclusions in Brazilian emerald. Note the yellower green colour compared to the Colombian emerald below. FOV 2.56 mm.

▶ A jagged three-phase inclusion inside a Colombian emerald, with a square salt (halite) crystal and round carbon dioxide bubble in liquid. FOV 0.91 mm.

Brazil is the third major source after Colombia and Zambia, commercially producing emerald since the 1960s with excellent finds of high-quality stones in the 1980s. Brazilian emerald may be lighter and more yellowish green than Colombian emerald. Other sources include Afghanistan, Pakistan (hosting the world's youngest at nine million years old) and Russia (discovered in the 1830s in schist in Malyshevo, Ural Mountains, along with other beryllium minerals, including chrysoberyl variety alexandrite and phenakite). It has also been found in China, India, Australia, the USA, Norway and Austria.

Emerald is typically included, in contrast to other beryl varieties, but due to its desirable colour and rarity, eye-visible inclusions are considered acceptable. A fine included emerald is many times the value of an equivalently sized flawless aquamarine, and extremely rare fine emerald with no eye-visible inclusions commands prices to rival diamond. The inclusions are described as the internal garden or 'jardin' – stones with few inclusions should be treated with suspicion. Inclusions are key to distinguishing natural versus

synthetic emerald, and to identify the host rock in which the emerald grew, which may indicate the source. Typical inclusions are growth planes, hexagonal zoning, solid mineral inclusions, partially rehealed fissures, and multiphase inclusions.

Zambia, Zimbabwe, Madagascar, Tanzania, Brazil and Russia host classic schist-type deposits. Emeralds may have rehealed fractures, and a roiled internal appearance caused by the erratic hydrothermal growth. Common inclusions indicating a schist host are phlogopite mica, and amphibole minerals including prismatic hornblende, jointed colourless actinolite crystals resembling bamboo, and hair-like tremolite, typical in Zimbabwean emerald. Orientated two-phase inclusions are common for Zambian and Brazilian emerald. Russian emerald has characteristic thin-film fluid inclusions which exhibit iridescence, orientated parallel to the basal plane.

Colombian emerald has a non-schist origin, containing inclusions indicative of the black shale and limestone hosts, such as rhombic calcite crystals and golden pyrite (iron sulphide). Three-phase inclusions of a liquid, a square crystal of halite (rock salt) and a carbon dioxide bubble are characteristic, caused by the hydrothermal fluids (brines) from which the emerald grew. This was considered diagnostic for many years, but is also seen in emerald from Nigeria, Afghanistan, Zambia and Ethiopia.

Similar to other beryl varieties, emerald may have thin, hollow, tubular inclusions parallel to the length of the crystal. When present in high enough quantity, and when the emerald is cut en cabochon, a cat's eye effect is revealed, or, on extremely rare occasions, a six-rayed star. Trapiche emerald is a rare, unusual find from Colombia. In cross section, these crystals have a cogwheel appearance with six spokes - the name 'trapiche' is the Spanish word for the wheels on Colombian sugar cane mills. Trapiche emerald has been known since the late 1800s. The central core is normally emerald on which are six growth sectors or arms of emerald, separated by zones of opaque black included material. These crystals are fashioned into polished slices or cabochons to reveal the structure and may have slight chatoyancy. How this intriguing emerald is formed is unknown.

Emerald is cut to emphasize the magnificent colour over clarity or brilliance. Cutting is a compromise between retaining carat weight, minimizing inclusions and maximizing the colour. The emerald cut is well-suited – it utilizes the elongate rough, and provides protection for the brittle corners. Lesser quality material may be fashioned into cabochons or beads.

Emerald is commonly treated to improve its appearance. Surface-reaching fractures are impregnated with oils or polymers/resin of a similar refractive index to reduce their visibility and improve clarity. These fillers may contain green dye to enhance pale emerald. It is thought more than 90% of emerald on the market is treated by fracture filling. Oiling, which is considered standard, requires general disclosure while fracture filling using dyes or resins requires specific disclosure. Oiling has occurred since ancient times. Cedarwood oil is commonly used today as its refractive index is similar to emerald. This treatment is not permanent as oil can dry out or leak over time, or turn yellowish, whitish or lumpy, requiring the gemstone to be reoiled. Polymers or epoxy resin are a more stable treatment but harder to remove if they decompose. Their refractive index is closer to emerald than oils, making fissures even less visible. Fracture fillings may be detected under magnification by a difference in surface lustre, trapped bubbles, flow structures, cloudy zones or dendritic dried areas. Light reflecting off the internal fissure interface gives a characteristic

▼ Closeup of a trapiche emerald, showing the six spokes and growth sectors. FOV 3.06 mm.

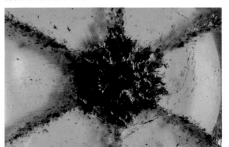

▼ Synthetic emerald faceted in a rectangular step cut with truncated corners, known as the emerald cut.

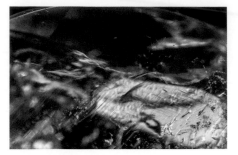

▲ Yellow, orange and blue flash effects from light reflecting off fracture fillings in an emerald gemstone. FOV 2.48 mm.

yellow, orange, pink or blue flash effect. Fracture fillings may also fluoresce white to yellow under longwave ultraviolet light. Dyes can be identified by colour concentrations in cracks. Treatment by heat and irradiation, commonly used on other beryl varieties, does not improve emerald's colour due to its chemical composition, so is not used.

This gemstone should be worn with care and is not recommended for everyday use. It is more brittle than other beryl varieties, and the inclusions and fractures make it less durable. If desired in an engagement ring, select an emerald with as few inclusions as possible. Avoid knocks, heat and sudden changes in temperature, especially if heavily included or fractured. As most natural emerald gemstones are fracture-filled, avoid chemicals and detergents that may dissolve and remove the filler. Remove emerald rings before washing hands or dishes. Wiping with a soft damp cloth is the safest method of cleaning, do not use ultrasonic or steam cleaners.

Emerald was first synthesized in the mid 1800s, but only produced commercially since the 1940s. All synthetic emerald should be disclosed and in the jewellery industry the terms laboratory-grown, laboratory-created, or [manufacturer name]-created are preferred. Synthetic emerald has far fewer inclusions than natural emerald, and when present the inclusions can indiate it was human-

made. There are two main methods of growth: flux melt and hydrothermal. Both use the ingredients aluminium, beryllium, and chromium metal oxides plus silica in the form of quartz or glass.

The flux melt method is the most common and uses a molten flux to dissolve the ingredients in a crucible of platinum or graphite. Through heat and pressure the ingredients dissolve and recrystallize, nucleating on a seed crystal of natural or synthetic emerald. Distinctive crystals grow slowly over eight to 12 months. Flux melt emerald has inclusions of twisted veils or smoke-like feathers filled with opaque undigested flux. Syngenetic crystals of the beryllium minerals phenakite and chrysoberyl may also be present as inclusions, growing at the same time as their host emerald. Chatham-created emerald, produced by the company Chatham, uses platinum crucibles sometimes causing inclusions of platinum platelets.

The hydrothermal method mimics the hydrothermal growth of natural emerald. A sealed container known as an autoclave is partially filled with water, the ingredients, and a suspended seed crystal. Under high temperature and pressure the ingredients dissolve and recrystallize on the seed crystal. This method is much quicker than flux melt. Growth features include chevron-like zones, pointed growth tubes and nail head spicules. While these also have a roiled appearance, there is no hexagonal zoning as seen in natural emerald.

The properties of synthetic and natural emerald differ slightly, but the differences are not consistent. Several methods of testing may be required to confidently distinguish between them, combined with a study of the inclusions. Occasionally advanced testing is required for positive identification. Synthetic emeralds may exhibit a strong red fluorescence, however some remain inert, like most natural emerald. Many, but not all, synthetic emeralds show a strong red appearance under the Gem-A Chelsea Colour Filter, while

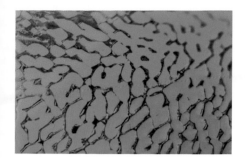

natural stones normally appear weakly red. Flux melt emerald typically has lower specific gravity and refractive index than natural emerald. Hydrothermal emerald has only slightly lower values, which may fall within the range of natural emerald.

Emerald may be used to create composite gemstones of an emerald crown on a base of glass or low-quality emerald. It may be simulated by green glass, yttrium aluminium garnet (YAG), cubic zirconia and aventurine quartz. Composite simulants include garnet-topped doublets of a green glass pavilion topped by a garnet crown.

◀ The partially healed fracture in this flux melt synthetic emerald contains whitish flux residue. FOV 1.09 mm.

▲ Nail-head spicules and roiled appearance inside a Biron-created hydrothermal synthetic emerald. FOV 1.88 mm.

Triplets of rock crystal and/or pale beryl forming the crown and pavilion, sandwiching a green layer, are known as soudé emerald. Emerald may be confused with grossular garnet variety tsavorite, chrome tourmaline and chrome diopside, but can be distinguished by different optical properties and its distinctive absorption spectrum, caused by chromium

Aquamarine

Composition $Be_3Al_2Si_6O_{18}$ beryllium aluminium silicate **Crystal system** hexagonal **Hardness** 7.5–8 **Cleavage** indistinct **Fracture** conchoidal **Lustre** vitreous **SG** 2.65–2.80 **RI** 1.563–1.596 **Birefringence** 0.004–0.010 **Optical nature** uniaxial - **Dispersion** 0.014

Aquamarine is the sky blue to sea green variety of beryl. It is named for its colour, derived from the Latin 'aqua marina' meaning water of the sea, and is traditionally said to bring luck to sailors. This popular gemstone has high durability, and is often found in large flawless crystals allowing for sizeable gemstones fashioned in a range of cutting styles. Water-clear gemstones over 15 ct are common, and gemstones in the hundreds of carats are known.

▲ A large, near flawless aquamarine gemstone from Russia, weighing 898.7 ct.

One of the most famous is the Dom Pedro, the largest known cut aquamarine, carved into an obelisk nearly 35 cm (13½ in) tall and weighing 10,363 ct (2.07 kg, 4.57 lbs). It was faceted in Idar-Oberstein, Germany, in 1992, from an original crystal that was around 1 m (3 ft) long, discovered in the late 1980s in Pedra Azul, Minas Gerais, Brazil. The crystal was accidently dropped, splitting into three pieces of which the largest became the Dom Pedro. It is now housed in the collections of the Smithsonian Institution in Washington, DC, USA. The Natural History Museum, London, holds a near flawless 898.7 ct faceted aquamarine gemstone.

The colour of aquamarine ranges from near colourless to light blue, through to deep blue, blue-green or green. The majority are of light tone. The most desirable is a medium blue to slightly greenish blue. Aquamarine is strongly dichroic exhibiting near colourless to pale blue/darker blue. The colour is caused by impurities of iron, and the colour range blends to the greens of heliodor.

Elongate hexagonal crystals are typical of aquamarine. Terminations are flat, sometimes with bevelled edges from pyramidal crystal faces, or tapered. Crystals may be striated along the length and are sometimes etched with beautiful surface patterns. The well-formed large sizes and clear nature make aquamarine specimens desirable to collectors.

Aquamarine predominantly forms in granitic pegmatites. Brazil has been the top producer since the early 1800s, and its finest crystals come from pegmatite and gem gravel deposits in the Minas Gerais state. There have been several significant finds. In 1910 the largest gem crystal found to date, in the world, was uncovered in the alluvial deposit at the Papamel Mine, Marambaia Valley, measuring 48.5 x 38 cm (19 x 15 in) and weighing 110.5 kg (244 lbs). It was a waterworn, doubly terminated crystal, described as flawless and so clear you could read through it. The majority of the crystal was blue-green and was estimated to produce 200,000 ct of gemstones. In the 1950s, the Santa Maria aquamarines were discovered at the Santa Maria de Itabira mine. Prized for their unusually saturated medium to dark blue, with no hints of yellow or green, they are the rarest and

◀ Elongate hexagonal aquamarine crystal of gem quality from Spitzkopje, Namibia.

▼ A dichroscope reveals the two pleochroic colours (dark/ pale blue), which combine to create the overall medium blue colour of this Vietnamese crystal.

Dichroscope

▶ Fancy cut 6.85 ct (left) and fantasy cut 9.23 ct (right) aquamarine gemstones from Madagascar.

most expensive aquamarine. This colour cannot be created through heat treatment. The name is now also applied to highly valued aquamarine of similar colour found on the African continent.

Madagascar provides fine blue gems and Pakistan is a significant producer. Other sources include Afghanistan, Kenya, Zambia, Nigeria, Mozambique, Namibia, Russia, the USA, Uruguay, India, Sri Lanka, Myanmar, China and Australia.

Most faceted aquamarine is inclusion-free. When present, hollow or liquid filled thin tubes orientated parallel to the length of the original crystal give the appearance of falling rain. This feature is so well known it is considered diagnostic. Other inclusions are wavy rehealed fractures, two- or three-phase inclusions, solid crystals of minerals such as mica, and spiky cavities. Chatoyant material creates attractive cat's eye aquamarine.

Due to the strong pleochroism, aquamarine rough is cut to show the deepest colour face up. Fortuitously, this is with the table facet cut parallel to the length of the crystal, allowing the best yield from the elongate rough. Step cuts are the prevalent cutting style, with deeper cut stones enhancing the colour. A well-proportioned gemstone, however, shows greater brilliance, and round or oval brilliant cuts are also common.

Aquamarine is popular for fantasy cut gemstones and carvings, particularly in larger sizes. This hard, durable gemstone is used in all forms of jewellery. It is safe to clean by ultrasonic or steam cleaners unless containing fluid inclusions or fractures. Its brittle nature, however, means it is vulnerable to knocks, and avoid heat which may fracture included stones.

Most aquamarine is heat treated to enhance the blue colour by removing green overtones, an undetectable treatment. Greenish, bluish-green, or even yellowish stones are heated producing a stable blue colour. Untreated aquamarine does not have a significant price difference to heat treated. Irradiation can produce the yellow colour of heliodor, and also an unstable blue or green colour. Rarely, surface-reaching fractures are filled. Synthetic aquamarine is known, but not often seen on the market.

Aquamarine may be confused with similarly coloured blue topaz and zircon, but is easily distinguished by a lower refractive index. It may be imitated by pale blue glass and synthetic spinel, neither of which exhibit pleochroism. The Gem-A Chelsea Colour Filter is a useful tool to quickly separate these blue gemstones, where aquamarine appears pale greenish-blue, topaz appears pale orange and glass and synthetic spinel appear red.

Heliodor

Composition Be$_3$Al$_2$Si$_6$O$_{18}$ beryllium aluminium silicate **Crystal system** hexagonal **Hardness** 7.5–8 **Cleavage** indistinct **Fracture** conchoidal **Lustre** vitreous **SG** 2.66–2.87 **RI** 1.562–1.602 **Birefringence** 0.005–0.009 **Optical nature** uniaxial - **Dispersion** 0.014

▲ Golden yellow step cut 3.48 ct heliodor gemstone.

The warm sunny colours of this beryl variety earned it its name heliodor, from the Greek 'doron' meaning gift and 'helios' for sun. Yellow beryl has been known for several centuries but it was not until the discovery in 1910 of significant quantities in Namibia that it was named heliodor. Heliodor ranges from greenish-yellowish to yellow to orangey yellow, coloured by impurities of iron. This variety is not well defined. Some deeper golden yellow colours are known as golden beryl, and light green caused by iron (not chromium) as green beryl, reserving heliodor for greenish-yellow colours only. However, these names are used interchangeably on the market, and are classed here as one variety. Yellow to greenish heliodor is dichroic lemon yellow/brownish yellow, with less distinct pleochroism than seen in other beryl varieties.

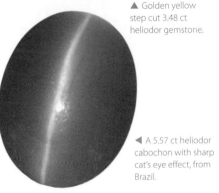

◀ A 5.57 ct heliodor cabochon with sharp cat's eye effect, from Brazil.

Heliodor forms elongated hexagonal prisms, often etched with rectangular pits in interesting patterns along the length. While less common than aquamarine in gem quality, heliodor can grow in large transparent crystals allowing the faceting of huge flawless gemstones. The Smithsonian Institution in the USA holds a flawless gem of 2,054 ct. Most faceted heliodor has relatively few visible inclusions, commonly two-phase liquid and gas inclusions in thin parallel tubes, negative crystals, rehealed fissures, and solid mineral crystals. More included material may be fashioned into cabochons, and, when abundant, the thin parallel tubes can create a sharp cat's eye effect.

Heliodor occurs in granitic pegmatites, and is found in many of the same locations as aquamarine. Brazil and Madagascar are dominant sources, along with the Ukraine, Russia, Namibia and the USA.

Heat treatment can remove the yellow colour and produce blue. As aquamarine is considered more valuable, naturally coloured heliodor is therefore less common on the market. Saturated golden yellow or lemony yellow colours may be created by irradiation of aquamarine and some iron-bearing goshenite. These treatments are impossible to detect, and much heliodor on the market has been produced by irradiation. All treatments should be disclosed. Synthetic heliodor is known but not often seen.

This gem may be confused with chrysoberyl, yellow apatite, topaz or citrine although the latter two commonly have a brownish or orange overtone. It may be distinguished by differing refractive indices.

Morganite

Composition Be$_3$Al$_2$Si$_6$O$_{18}$ beryllium aluminium silicate **Crystal system** hexagonal **Hardness** 7.5–8 **Cleavage** indistinct **Fracture** conchoidal **Lustre** vitreous **SG** 2.71–2.90 **RI** 1.572–1.602 **Birefringence** 0.005–0.009 **Optical nature** uniaxial - **Dispersion** 0.014

▶ Morganite crystal, 8.8 cm (3.46 in) wide from Urucum Mine, Minas Gerais, Brazil.

▼ The 598.7ct flawless morganite gemstone from Madagascar, acquired in 1913, only three years after morganite was first discovered.

Morganite is the light pink to orange-pink variety of beryl, ranging from peach to salmon pink to bubble gum pink. The often delicate hues are caused by impurities of manganese. Saturated pinks are the most desirable, and the majority of morganite is treated to achieve this. Interestingly, this has led to the rise in popularity of salmon pink morganite through the desire to own untreated gems. First found in Madagascar in 1910, morganite was named after the famous American financier J P Morgan, an important gem collector of the time. Although found in the same pegmatitic environments, morganite is rarer than aquamarine and heliodor. It can form zoned crystals with aquamarine or heliodor, indicating a change in the geological environment (and therefore the incorporated impurities) as the crystal grew. Madagascar remains an important source of beautiful rose-coloured crystals. Other sources include Brazil, Afghanistan, Pakistan, Mozambique, Namibia, Zimbabwe, Myanmar, Sri Lanka (gem gravels), and the USA.

Like all beryl, morganite forms hexagonal prisms, favouring short prismatic to tabular crystals compared to elongate aquamarine and heliodor. The terminations are typically flat, often with bevelled edges. Crystals are commonly incomplete. Morganite can grow to large sizes, with crystals heavier than 10 kg (22 lbs) known, and faceted gemstones come in a range of sizes. The Natural History Museum, London, holds an exceptional 598.7 ct bubble gum pink gemstone from Madagascar, thought to be the largest flawless, faceted morganite in the world

This mineral is transparent to opaque, with faceted gemstones normally eye clean. More included material may be fashioned into cabochons or carved. Fluid inclusions are common, and often multiphase. Morganite exhibits distinct dichroism of bluish-pink/pale pink so cutters orientate the rough to obtain the best colour face up. Most morganite is pale, so is often cut with a deep pavilion to enhance the colour. Larger gemstones will have a greater depth of colour. Step and mixed cuts are common, with the larger sizes allowing for fancy cuts. Morganite is inert or fluoresces weakly under UV light with a pinkish or violet reaction.

The colour is commonly improved by heat treatment. Salmon pinks are heated to remove orange and yellow tints, and deepen the pure pink. This treatment is not detectable and the resulting colour is stable. Morganite may also be irradiated to improve the pink colour, however both irradiated and untreated morganite can fade slowly over time in strong light. Goshenite that contains manganese may be irradiated to produce pink, similarly undetectable.

Synthetic morganite grown via the hydrothermal process is seen on the market, and should be disclosed as such. Morganite may be confused with spodumene variety kunzite, pink topaz, rose quartz, pink apatite and scapolite. It can be distinguished by its fluorescence and differing optical properties.

Red beryl

Composition $Be_3Al_2Si_6O_{18}$ beryllium aluminium silicate **Crystal system** hexagonal **Hardness** 7.5–8 **Cleavage** indistinct **Fracture** conchoidal **Lustre** vitreous **SG** 2.66–2.87 **RI** 1.560–1.577 **Birefringence** 0.004–0.008 **Optical nature** uniaxial - **Dispersion** 0.014

▲ A 1.19ct gemstone with a 1.7 cm (0.66 in) hexagonal crystal of red beryl from Wah Wah Mts, Utah, USA.

Red beryl is the rarest of all beryl and very rare in gem quality. This deep red to raspberry red variety is known from very few locations, only producing notable mineral specimens and gem rough in Utah, the USA. It was first discovered in 1904 in the Thomas Range on a prospecting claim for topaz. A second find was made in the Wah Wah Mountains in 1958, later becoming the Ruby Violet mine. This is the only mine to produce significant quantities of gem material, however due to the low yield the mine is no longer producing. Red beryl is always found in the volcanic rock rhyolite, forming from late-stage fluids reacting with beryllium contained in the rhyolite.

Small hexagonal prisms are typical of red beryl, with flat terminations often at both ends. Mineral specimens of these lovely crystals, set against a backdrop of pale rhyolite, are highly sought by collectors. Crystals are generally 1 cm (0.4 in) or less, thus faceted gemstones are small weighing less than 1 ct, however larger gemstones of several carats do exist.

The distinctive deep red to pink colour is caused by impurities of manganese. Red beryl also contains iron and is inert under UV light. It is typically included, ranging from opaque to translucent, very rarely transparent. Fractures, solid minerals, and wavy fluid or fingerprint inclusions are common. Most red beryl is fracture-filled with resin or oil to improve the clarity, and high-quality unoiled gemstones command high prices. Synthetic red beryl is known on the market and can be distinguished by the lack of natural inclusions and internal chevron-like zoning.

Red beryl was originally named bixbite, after the prospector Maynard Bixby who discovered it. However, due to the closeness of the name to the mineral bixbyite (also discovered by Bixby) it is no longer used. Misleading names such as red emerald should be discouraged.

Chrysoberyl

Composition $BeAl_2O_4$ beryllium aluminium oxide
Crystal system orthorhombic **Hardness** 8.5
Cleavage good **Fracture** uneven **Lustre** vitreous
SG 3.68–3.82 **RI** 1.739–1.777 **Birefringence**
0.007–0.013 **Optical nature** biaxial +
Dispersion 0.015

This transparent yellow to green mineral is both attractive and durable, so is well suited for use as a gemstone. The name chrysoberyl is derived from a combination of the Greek 'chryso' meaning golden, and 'beryllos' for the beryllium content. It was known as chrysolite in the 1800s to the early 1900s, the same name used for other golden gemstones such as peridot, but this historical name is no longer in use, avoiding confusion.

Chrysoberyl has two rare varieties, cymophane and alexandrite, both with defining optical phenomenon. Cymophane is fittingly named, derived from the Greek for waving light, because it exhibits a chatoyant cat's eye with a band of light that moves over its surface. This variety may also be known simply as cat's eye and is the only chatoyant gem to have this privilege without a mineral designation. Alexandrite is a rare and highly-sought after variety of chrysoberyl, which exhibits a colour change from green to red. First found in Russia, it was reputedly named for Czar Alexander II.

This gem is a beryllium aluminium oxide, containing the rare element beryllium and requiring high enough concentrations in order to grow. It is therefore formed through pegmatitic processes where these rare trace elements are enriched in the late stages of crystallization of large igneous intrusions. Chrysoberyl may occur in pegmatites, or in the surrounding host rocks where beryllium- and aluminium-rich fluids interact with the host rock during contact metamorphism. Due to its high durability, chrysoberyl is weathered and transported from its host, and found as waterworn pebbles in gem gravel placer deposits, along with other gem minerals such as corundum and spinel.

▶ A chrysoberyl trilling with three v-shaped twins in cyclic form from Espírito Santo, Brazil.

▼ A 57.1 ct modified brilliant cut chrysoberyl gemstone.

The main sources of chrysoberyl are Brazil, Sri Lanka and Myanmar, commonly mined from placer deposits. Other sources include Zimbabwe, Tanzania, Madagascar, India, Pakistan, Russia, Australia and the USA.

Part of the orthorhombic crystal system, chrysoberyl favours tabular or short prismatic crystals. Twinning is common with V-shaped penetration twins. These may be repeated in a three-fold cyclic form with the twins at 120 degrees and a pseudo-hexagonal outline. This is known as a trilling. Twins are defined by striations on the crystal faces and re-entrant angles between faces.

Chrysoberyl forms yellow, golden, greenish yellow to greenish brown, and green coloured crystals. Golden yellow or greenish yellow are the most desirable. Colours are typically pale, with deeper colours attaining higher value. The colour is caused by small amounts of iron impurities, and chrysoberyl does not normally fluoresce due to this iron content. Chrysoberyl is pleochroic, a phenomenon seen most strongly in brown stones showing brown and brownish yellow when viewed from different directions.

▶ 'The Hope Chrysoberyl', a flawless 44.94 ct gemstone, once in the collection of Henry Philip Hope, now held in the Natural History Museum, London.

While many natural crystals are opaque or heavily included, gem-quality chrysoberyl is transparent and often eye clean. Common inclusions are fingerprints and feathers. The repeated twinning may be seen internally as parallel lines or stepped planes. Elongated tubes or needles of minerals such as rutile are common.

Chrysoberyl is normally faceted, most commonly into brilliant and step cuts, in a variety of shapes. Large transparent gemstones are known but most are less than a few carats. Chrysoberyl has a bright vitreous lustre, owing to its high refractive index, forming brilliant stones. However, due to its low dispersion it does not have much fire.

It is a very durable gem with a high hardness at 8.5 (one of the hardest natural gemstones after diamond and corundum), fair to good cleavage, and excellent toughness. It is also a very stable mineral, resistant to chemical attack. This, combined with its transparency and high lustre, makes it a beautiful and versatile gemstone, suitable to wear every day in all types of jewellery. It is safe for most cleaning methods, however it is not recommended to use heat, such as a jeweller's torch, on cat's eye material.

Chrysoberyl is generally not treated. It may be confused with many yellow or green gems including beryl, topaz, andalusite, peridot, sapphire, tourmaline, garnet and citrine, but is distinguished by a differing refactive index.

Cymophane

A small portion of chrysoberyl will show a chatoyant cat's eye effect and is known as cymophane. These specimens similarly range from yellow or golden, to greenish brown or grey, coloured by impurities of iron. The chatoyancy is caused by multiple parallel needle-like or tubular inclusions. These are usually so tiny that the stone appears transparent to translucent. When cut en cabochon with the base parallel to the inclusions, the cat's eye phenomenon is revealed. The highest valued are 'milk and honey' colour, a translucent pale yellow brown stone with a sharp white band of reflected light. Chrysoberyl cat's eye is the finest among all gems. Natural cat's eyes are normally cut as a double cabochon, retaining more inclusions to enhance the effect, and also increasing the weight.

Cymophane is found in most locations for chrysoberyl and is best known from Sri Lanka and Brazil. Other sources include China, Tanzania, India, Zimbabwe and Myanmar. Cymophane may be imitated by cat's eye quartz, cat's eye tourmaline, cat's eye scapolite, bleached tiger's eye, cat's eye opal, and even human-made fibre optic glass. It may be distinguished by its harder nature.

▼ A 5.28 ct 'milk and honey' chrysoberyl cat's eye.

Alexandrite

Alexandrite is the rare variety of chrysoberyl that shows a colour change effect from green to red depending on the light source in which it is viewed.

It has the same chemical composition as chrysoberyl but is coloured by impurities of chromium. Chromium and beryllium do not normally occur together in nature, so this mineral is only found in a few locations around the world. Alexandrite was first discovered in the Ural Mountains in Russia following the opening of the emerald mines in the early 1830s, said to be named after Czar Alexander II. It is highly valued, as red and green are the imperial colours of Russia.

Crystals are commonly opaque or very included with poor clarity and many fractures, so are unsuitable for use as gemstones. Gem quality is rare, with faceted gemstones over 2 ct extremely rare. Superb transparent gemstones over 1 ct can rival the finest ruby, sapphire and emerald. Due to its scarcity, and rarity in fine quality and large size, alexandrite is a highly sought after gemstone. Even rarer still is cat's eye alexandrite, with the chatoyancy of cymophane and a colour change.

▲ A fine crystal group of alexandrite with trillings, from Tokovaya mine, Malyshevo, Ural Mountains, Russia, 13 cm (5 in) wide.

The colour change effect results from alexandrite's ability to transmit visible light in both red and blue-green wavelengths. The colour it appears is determined by the lighting in which it is viewed. In daylight or cool lights, which are richer in blue and green wavelengths, it appears green, bluish green or yellowish green. This is because the human eye detects green wavelengths better than red. In warm lights, such as incandescent or candlelight, which contain more wavelengths of red, it is enough to tip the balance and the alexandrite appears red, purple or purplish brown.

▲ A 17.25 ct alexandrite with many inclusions from Malyshevo, Ural Mountains, Russia, appearing bluish green in daylight (left) and purplish red under incandescent light (right).

The most desirable colour change is from vivid green to intense raspberry red, described as 'emerald by day, ruby by night'. The effect is so remarkable that this colour change phenomenon in any gem is known as the alexandrite effect. Alexandrite also has strong pleochroism, appearing green/orange/purple-red when viewed in different directions. This optical property is independent of the colour change effect, however the colour change is strongest when viewed in one of the pleochroic directions. Alexandrite may show weak red fluorescence under longwave and shortwave UV light, due to its chromium content.

Many of the finest alexandrites came from the mica schists of the Ural Mountains in Russia, although these historical mines produce little today. Brazil is now the main source of fine specimens. Other important sources include Myanmar, Sri Lanka, Tanzania, Madagascar, India, Zimbabwe and Zambia.

Gem-quality alexandrite is cut and faceted to show the best colour change when viewed face up. A gem cutter must also take into account the orientation of the pleochroism in the rough. Mixed cuts with a brilliant cut crown and step cut pavilion are the most common and tend to be oval or cushion shapes. Chatoyant material is cut en cabochon to reveal the cat's eye effect.

Alexandrite is not usually treated. Surface reaching fractures may be filled to improve clarity and durability. It is safe to clean in warm soapy water. Avoid ultrasonic and steam cleaners, particularly if heavily included or fracture-filled. It is also recommended to avoid knocks and extreme heat.

Synthetic alexandrite has been known since the 1970s, exhibiting a good colour change. The most commonly encountered is grown by the crystal-pulling (Czochralski) method, showing purplish red to blueish. The stones are eye clean but may have internal curved striations seen under magnification, and are distinguished from natural alexandrite by the difference in the

▼ A 43.18 ct Sri Lankan alexandrite with few flaws, appearing dark green in daylight (top) and dark brownish red in incandescent light (below).

inclusions. Other synthetics are grown by flux melt or floating-zone techniques, with a purplish-red to green colour change. Prior to the production of high-quality synthetic alexandrite, this gem was frequently imitated by vanadium-doped synthetic corundum. Confusingly, this was often sold as 'synthetic alexandrite'. This simulant shows the characteristic curved zoning of flame-fusion synthetics, and a colour change from purple-pink to slate blue, with an appearance closer to that of amethyst. Colour change synthetic spinel (green to red), colour change garnet (bluish-green to purplish-red), and colour change glass containing neodymium (bluish-green to purplish-pink), have also been used as imitations.

Corundum – ruby and sapphire

Ruby and sapphire are two of the most treasured gems, prized for millennia for their rich colour, bright lustre, and durability. What surprises many is that they are the same mineral – corundum. Ruby, the king of precious stones, is the red gem variety, while sapphire is all other colours, most famously blue. Ruby and sapphire, together with beryl variety emerald, are the 'big three' in coloured stones.

Corundum species

Composition Al_2O_3 aluminium oxide **Crystal system** trigonal **Hardness** 9 **Cleavage** none **Fracture** uneven, conchoidal **Lustre** adamantine, vitreous **SG** 3.90–4.10 **RI** 1.756–1.780 **Birefringence** 0.008–0.010 **Optical nature** uniaxial - **Dispersion** 0.018

▼ 'The Edwardes Ruby', a 162 ct tabular ruby crystal with triangular markings, thought to be from Myanmar.

Ruby and sapphire are popular gems, thought to account for more than 50% of global coloured stones production. This is in part owing to greater availability in the past few decades, due to new sources and treatments. The name corundum likely originates from the Sanskrit 'kuruvinda' meaning ruby. It was not until the late 1700s, with advances in scientific knowledge, that ruby and sapphire were determined to be the same mineral, and distinguished from other red and blue gems.

Corundum is an aluminium oxide. When pure it is colourless, however most contains impurities such as iron or chromium which substitute for aluminium in the crystal structure. Corundum has a trigonal crystal symmetry, typically forming hexagonal prisms, but the crystal shape varies with colour and source. Ruby usually has flat terminations and may be short and tabular. Sapphire is commonly barrel-shaped, or elongated bipyramidal. Crystal faces are rarely flat but usually uneven or stepped. Pyramidal terminations are striated perpendicular to the length, and flat terminations show triangular striations and steps. Crystals have greasy to dull

◄ An elongate bipyramidal crystal of blue sapphire with striations perpendicular to the length.

lustre. Polysynthetic, or repeated, twinning is commonly seen on crystal faces as striae in two directions at almost 90 degrees.

Corundum forms in aluminium-rich, silicon-poor geological environments. Because silica (SiO_2) is prevalent in Earth's crust, corundum is relatively uncommon, and gem quality rare. Even so, corundum is found all over the world, mined from both primary and secondary deposits.

The formation of corundum requires high temperature and medium pressure conditions. This is linked to large-scale tectonic events including continental collision with mountain building, and volcanism associated with continental rifting. These events are also responsible for bringing corundum to the surface. Primary corundum deposits therefore occur in metamorphic rocks such as marble (metamorphosed limestones/dolomites), amphibolite (metamorphosed mafic rocks), skarns (formed by metasomatism), gneiss and schist, as well as in igneous rocks primarily alkali basalt, syenite and lamprophyre dykes.

Research shows three main periods of corundum formation. The first was during the Pan-African Orogeny (750–450 million years ago) when the supercontinent Gondwana formed. The collision of continents formed the Mozambique Orogenic Zone, creating conditions favourable for the formation of gem minerals. Ruby and sapphire deposits in Sri Lanka, southern India, Madagascar and the gemstone belt of East Africa (Kenya, Tanzania) lie within this zone, formed between 600 and 500 million years ago during peak metamorphism. The second period was during the collision of the Eurasian and Indian tectonic plates, which pushed up the Alps and Himalayas (45–5 million years ago).

▼ An etched crystal of deep red translucent ruby, unusually still attached to its marble matrix, from Mogok, Myanmar.

Compressional processes formed metamorphic belts in central and southeast Asia creating the marble-hosted ruby deposits through Afghanistan, Pakistan, and Myanmar to Vietnam. The third period relates to volcanic extrusion of alkali basalts (65–1 million years ago) in areas of crustal extension. Gem corundum, having crystallized under magmatic and/or metamorphic conditions deep in Earth's crust, was brought to the surface within basaltic magma. These deposits occur in Australia, southeast Asia, Nigeria, Cameroon, Ethiopia and Madagascar.

Due to its high hardness, durability and density, corundum can survive weathering and transportation. It accumulates over millions of years into secondary placer deposits such as alluvial gem gravels, found with other hardy gems including spinel, beryl and chrysoberyl. Most corundum is mined from placer deposits, which is easier and less expensive than mining hard rock primary deposits, and the corundum is already naturally sorted and concentrated. Most mining is small scale or artisanal, usually by hand with some mechanised mining, digging down to gem-bearing layers and sifting through sediment.

The geographic origin of a ruby or sapphire has a dramatic effect on its value, influenced by the prestige, and the ethical, environmental and political concerns. Untreated gemstones also attract high premiums, due to the prevalence of treatment. As a result, the identification of gem origin and treatment is a driving force in gemmological research. The study of inclusions can indicate the parent rock and source location, or synthetic origin, as well as provide evidence of any treatments. Detailed analytical methods, such as trace element geochemistry, spectroscopic analysis and determination of oxygen isotopes, can indicate metamorphic versus magmatic origin and potential geographic source. High value gemstones are usually evaluated by one or more gemmological laboratories to verify natural origin and lack of treatment.

Ruby and sapphire have many similar inclusions:

- Silk, the most prevalent, is a fine haze of minute crystals. These appear as whitish particles, long thin needles of titanium oxides such as rutile (TiO_2), and tiny platelets of iron oxides such as hematite (Fe_2O_3) and ilmenite ($FeTiO_3$). These are orientated in three directions at 60 degrees in the basal plane, following the underlying trigonal symmetry of the host corundum. Very fine silk is cloud-like, while larger needles form reticulated grids. Silk is thought to form from impurities within the corundum that exsolved as the host cooled.

- Straight angular growth zoning, commonly hexagonal, defined by colour zoning, growth lines or zones of silk.

- Crystals of other minerals including spinel, apatite, calcite, feldspar, mica, and zircons with dark haloes (circular stress fractures).

- Negative crystals (tiny internal voids following the shape of the host crystal) which may be liquid- and/or gas-filled.

- Polysynthetic (repeated) twinning, seen as parallel striations.

- Feathers or fingerprints from partially healed fractures. Tiny cavities or liquid inclusions remain along the fracture plane in wispy, veil-like or fingerprint patterns.

▶ A rare 12-rayed star caused by silk inclusions of rutile (gold arms) and hematite (silver arms). Hexagonal growth zoning is also seen.

When silk occurs in large enough quantities, it produces asterism, seen when the ruby or sapphire is cut en cabochon. As silk crystallizes in three directions, it creates a six-rayed star. On rare occasions a 12-rayed star is produced, most commonly in black sapphire. This occurs when both rutile and hematite-ilmenite silk is present, as they crystallize offset by 30 degrees. The value of star stones is determined by the saturation of colour, sharpness of star, and transparency of the stone. The most famous is the Star of India, a 536.35 ct blue stone from Sri Lanka, housed in the American Museum of Natural History, New York. Larger star stones are known.

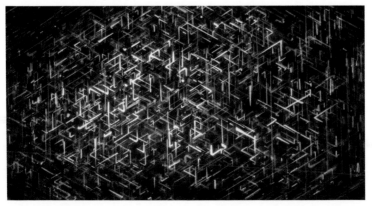

◀ Inclusions of iridescent silk in a Vietnamese ruby, orientated in three directions at 60 degrees. FOV 1.84 mm.

▼ Corundum is cut in different styles - mixed cut ruby (left) and step cut purple sapphire (right).

Corundum is second only to diamond in hardness, and is the reference mineral for 9 on the Mohs scale. The hardness varies in different directions across the crystal, which must be considered when polishing. It has no cleavage, however it can be brittle, and has a parting in three directions along planes in which inclusions lie, one of which is the basal plane. Ruby and sapphire take a good polish due to the hardness, with vitreous to subadamantine lustre. Corundum is pleochroic, showing two colours when viewed from different directions. It has a high refractive index creating brilliant stones, however with low dispersion there is little fire, but this is more than made up for by the stunning colours.

The hardness and toughness give non-gem quality, and synthetic corundum many industrial uses. These include as abrasives (emery is corundum with a mix of iron oxides and other impurities), bearings in watches, scratch-resistant optics, and in

electronic and scientific instruments, for instance ruby was the first material used in lasers.

As high value gems, ruby and sapphire are cut to retain maximum weight while displaying the best colour. Cutters must account for the pleochroism, orientating the rough to show the desired colour. A range of styles are used with mixed cuts common for strongly coloured material, while pale or colourless stones may be brilliant cut. Oval and cushion shapes are prevalent. Lower quality material is fashioned into beads or carved as gems or ornamental objects. Corundum with asterism is cut en cabochon. Most gem-quality corundum is small, with fine rubies rare over 3 ct, and fine blue sapphires rare over 5 ct.

Most ruby and sapphire on the market is treated. Heat is used to improve colour and clarity, and to heal fissures. As corundum forms under high

▶ Discoidal tension fractures around negative crystal inclusions indicating heat treatment of this sapphire. FOV 2.47 mm.

◀ A colourless synthetic sapphire boule, 5.5 cm (2 in) long, split longitudinally to relieve internal stress.

▼ A 4.95 ct synthetic ruby cabochon exhibiting asterism with a central sharp star.

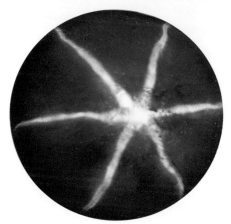

temperature, it can withstand heating to 2,000°C. Its inclusions, however, have lower melting points. Good indicators of heat treatment are solid mineral inclusions with a melted snowball appearance, or signs of stress like discoidal tension cracks from differential expansion. Minerals forming silk are resorbed leaving skeletal silk. Gemstones with intact natural silk indicate they have not been heated to high temperatures. Other treatments such as fracture filling and diffusion are occasionally seen.

Corundum has a simple chemistry (Al_2O_3) so is easy to synthesize, and ruby was the first major gemstone to be manufactured. Although first synthesized in the 1800s, commercial production did not begin until after 1902, when French chemist Auguste Verneuil published the flame-fusion method. This grows a single rounded crystal called a boule, very cheaply and quickly in a few hours. Synthetic star stones, first made in 1947, are also created by flame-fusion by adding rutile to the alumina powder feed. Cabochons with very sharp, even, central stars, and flat polished backs are suspicious. Natural star stones often have curved backs to increase carat weight. Other, lesser used methods of synthesis include flux melt (around one year to grow), hydrothermal (a few months), and rarely Czochralski crystal-pulling. A range of colours can be created by doping with various impurities.

Synthetic corundum tends to be eye clean, with no natural inclusions but other distinctive ones.

Flame-fusion gems have curved, never straight, growth lines as striae or colour zoning, sometimes challenging to spot. Rounded or elongated gas bubbles are common, with fine clouds dispersed along curves. It is worth noting that natural looking flaws such as fingerprints can be artificially induced. Flux melt and hydrothermal rubies grow as crystals with faces, so have straight angular zoning. Flux melt gems lack natural solid or liquid inclusions, but many contain the seed crystal on which they grew, or metallic platelets from the crucible. Undissolved flux appears as breadcrumbs, comets or feather inclusions resembling natural liquid fingerprints, with a characteristically twisted veil-like or wispy appearance, similar to those in flux-grown emerald. The crystal-pulling method creates clean crystals with few inclusions of elongated bubbles.

As synthetic corundum is of lower value, less care is often taken during faceting. The polish and finish tend to be of lower quality, and overheating by fast polishing creates chatter marks of small wavy parallel cracks on the facet faces. This may also be seen on low-quality natural gemstones.

Ruby

Composition Al$_2$O$_3$ aluminium oxide **Crystal system** trigonal **Hardness** 9 **Cleavage** none **Fracture** uneven, conchoidal **Lustre** adamantine, vitreous **SG** 3.90–4.10 **RI** 1.756–1.780 **Birefringence** 0.008–0.010 **Optical nature** uniaxial - **Dispersion** 0.018

Ruby is one of the most coveted gemstones, with a bright red colour associated with power, wealth, love and desire. The name ruby has been used since medieval times, and stems from the Latin 'ruber' for red. Prior to this, anthrax and carbunculus were terms used for red gems, interpreted as ruby. However ruby is uncommon in ancient artefacts, probably due to its rarity and difficulty to work owing to high hardness, and many 'rubies' have since been identified as garnet. Ruby was also assumed to be the same mineral as red spinel, understandable as they are found together in marble and placer deposits, with similar appearance and properties. In fact many large treasured 'rubies', including the Black Prince's Ruby in the British Crown Jewels, are actually spinel.

This variety of corundum ranges from vibrant red to purplish or orangey red in medium to dark shades. The most desired colour is intense red with a hint of blue, traditionally known in Myanmar as pigeon's blood red. Light red stones are often termed pink sapphire, although there is no official colour boundary. Ruby has strong pleochroism of purplish red/orangey red.

Ruby's rich red colour comes from chromium, the element responsible for emerald's vivid green, required in quantities of at least 1%. Chromium also causes red fluorescence, so strong that it enhances the colour in daylight with a red glow. The presence of iron impurities however quench the fluorescence, diminishing vividness and imparting a brownish tinge. Ruby is less common than sapphire, as chromium is not as prevalent as other impurities. Gem-quality ruby is much rarer, especially in larger sizes, and most faceted gemstones are under 2 ct. Colour is a key factor in determining value, and the finest red rubies command some of the highest prices among all gemstones, at times reaching over US$ 1 million per carat.

The chromium content gives ruby a distinctive absorption spectrum, absorbing violet and green light, transmitting red and blue with a doublet (two close absorption lines) in the red. This spectrum is shown to some extent by chromium-bearing pink, purple and the unique orange-pink sapphires known as padparadscha. In some rubies the doublets are emission lines rather than absorption, indicating the wavelengths of the red fluorescence. This emission is why ruby is used for lasers.

▼ Brilliant cut ruby with glowing 'pigeon's blood red' colour, 1.15 ct, from Myanmar.

▶ A superb pinkish red natural star ruby, set in a ring with 20 diamonds.

It is incredibly rare for ruby to be flawless. The visibility of inclusions influences the price, and while it generally lowers the value, small quantities of silk can actually improve the colour by scattering the light, creating a brighter, redder gemstone. Ruby commonly contains fissures or fractures, more so than sapphire. Star ruby is relatively well known.

Unusually, ruby may show a trapiche pattern of six growth sectors bound by heavily included zones like spokes of a wheel, similar to trapiche emerald. It is known from several locations including Myanmar, Vietnam, Tajikistan, Pakistan and Nepal.

Ruby occurs in metamorphic rocks, hosted predominantly in marble and amphibolite, also associated with igneous alkali basalts. The most famous source is the Mogok Stone Tract in Myanmar, providing the finest rubies for centuries. Most are mined from gem gravel placer deposits with other gem minerals, while others are extracted from the marble host, which is challenging to mine. Burmese ruby is high in chromium with intense pigeon's blood red colour, and strong fluorescence. Mogok also produces fine star ruby. Ruby occurs elsewhere in Myanmar, notably the large deposits discovered at Mong Hsu in 1992. These rubies have purple cores and are heavily fractured, requiring heat treatment to remove the purplish colour, and fracture filling.

Sri Lanka produces fine ruby with a pinkish or purplish hue, which has strong fluorescence. It is well known for star ruby. Ruby is mined from extensive gem gravel deposits known as illam in the Ratnapura and Elahera areas. Their parent rock is metamorphic.

Thailand was an important source, and to a lesser extent Cambodia, with deposits along their border in the Chanthaburi-Trat and Palin regions. Here ruby (and sapphire) are associated with alkali basalts. The ruby is high in iron that quenches fluorescence, so it is a less vivid brownish or purplish red. It also lacks light-scattering silk, with a darker appearance. Today production is limited, and Thailand is better known for ruby cutting, treating and trading. Vietnam has been a major ruby producer since the 1980s. Marble-hosted primary deposits and secondary placer deposits occur in the north (Luc Yen, Yen Bai, and Quy Chau areas) and the rubies often have slightly pinkish to purplish overtones.

The discovery of the Montepuez deposit in Mozambique in 2009 changed the ruby market, with huge reserves of high-quality ruby creating a stable supply and putting Mozambique at the forefront. The main source of gem-quality ruby is secondary colluvial deposits, but primary amphibolite-hosted deposits are also mined. Other East African amphibolite-hosted deposits are found in Kenya, Malawi and Tanzania (Winza, Longido). Longido produces large opaque rubies in green zoisite rock called anyolite or ruby-in-zoisite. This was discovered in 1954 and is used for carving and decorative stone. Tanzania also

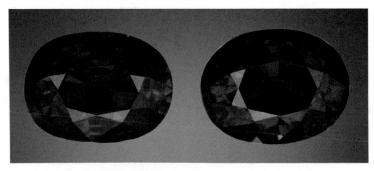

◀ Two matching unheated Madagascan rubies, mixed cuts with combined weight of 2.66 ct.

has marble-hosted ruby at Mahenge, Morogoro. Central eastern Madagascar produces fine rubies including pigeon's blood red.

Ruby is found in metamorphic marbles in Tajikistan, Afghanistan (Jegdalek, a historic source), Pakistan (Hunza Valley, fine colour but flawed), and Nepal. Amphibolite-hosted ruby is found in the USA (North Carolina), Australia (Northern Territory), and Japan, and it is associated with alkali basalt in Australia (Queensland, New South Wales). Other deposits include those in China, Norway, Macedonia and India (Mysore, including stars). Greenland boasts the oldest ruby deposits of 2.97 billion years old, formed through metasomatism.

This gemstone is commonly treated to improve colour and clarity, and treatments should be disclosed. Low temperature heat treatment can remove bluish colours, and will not disrupt inclusions, making it hard to detect. Heating to high temperatures will dissolve silk to improve clarity and remove purplish or brownish hues. Surface-reaching fractures are impregnated with oil, polymer or glass to improve the clarity, and these fillers often contain red coloration to improve hue. Lead glass has a similar refractive index to ruby making the filled fractures almost invisible. The glass can be detected by bubble inclusions, a crazed surface with different surface lustre, and light reflecting off the internal fissure interface may give a blue flash effect. Many low-quality rubies contain such a high percentage of lead glass filling that they are termed composite ruby rather than treated ruby.

Thin fractures may be healed by heating ruby in a flux such as borax, enabling the inside surfaces to dissolve and recrystallize shut. This is a common treatment of heavily fractured Mong Hsu ruby and is detected by glassy residual flux and bubbles trapped in the healed fissure. Some gemmologists consider this recrystallized ruby as synthetic, and it must be disclosed in addition to heat treatment.

▲ Blue flash effect from light reflecting off glass filled fractures in a natural ruby. FOV 2.90 mm.

▼ Emerald cut synthetic ruby gemstone.

Ruby is treated by diffusion, where gems are heated to very high temperatures in the presence of an element that diffuses into the crystal structure. Surface diffusion of titanium increases rutile silk, improving asterism. Lattice diffusion of beryllium creates strong reds in purplish or brownish stones that do not respond to heat treatment. These treatments require disclosure.

Ruby is durable, so cleaning with warm soapy water is safe, and ultrasonic and steam cleaners are generally safe unless the ruby is fracture filled or dyed. Extra care should be taken with lead-glass-filled composite ruby as it has very low durability, and even mild chemicals or acids such as lemon juice can damage the filling.

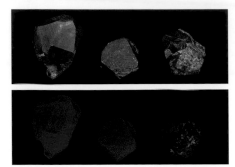

▲ Left to right: pure synthetic ruby, chromium rich Myanmar ruby and higher iron Tanzanian ruby under natural light (top) and longwave UV (bottom), showing a decreasing fluorescence response.

While most natural ruby contains some amount of iron, synthetic ruby can be pure, making its colour more vivid and the fluorescence stronger. Beware that manufacturers may add iron to reduce fluorescence, so the gem appears more natural. A combination of features, including inclusions, should be studied to distinguish synthetics.

As a valuable gemstone, ruby has many imitations. Red glass and plastic are distinguished by bubble and swirl inclusions and their amorphous nature. Red-dyed crackled quartz can be distinguished by different refractive index. Composite stones include garnet-topped doublets with an almandine crown on a glass base, distinguished by the refractive index of garnet and iron absorption spectrum. More difficult to detect is a doublet with a natural sapphire crown on a synthetic ruby base, as it has natural ruby's colour, inclusions and refractive index. These may be distinguished by the join seen as a line from the side, and by flattened gas bubbles trapped in the join.

Ruby's characteristic chromium absorption spectrum can be used to distinguish it from many red gems. Spinel and garnet also differ by their isotropic nature, and garnet and red tourmaline lack fluorescence. Beware of misleading trade names such as Bohemian Ruby which is garnet.

Sapphire

Composition Al_2O_3 aluminium oxide **Crystal system** trigonal **Hardness** 9 **Cleavage** none **Fracture** uneven, conchoidal **Lustre** adamantine, vitreous **SG** 3.90–4.10 **RI** 1.757–1.780 **Birefringence** 0.008–0.010 **Optical nature** uniaxial - **Dispersion** 0.018

Most people think of sapphire as blue, but it is all colours except red. Sapphire has been revered for millennia, believed to represent honesty, trust and loyalty, and worn by ancient Romans and Greeks for protection. It is the second choice after diamond for engagement rings, a perfect alternative for its colours, beauty and durability. The name is derived from the Greek 'sappheiros' meaning bright blue stone, although in ancient times this name referred to lapis lazuli. Sapphire is still associated with blue, and when used on its own the name refers to blue. Other colours were given names such as oriental amethyst for purple, and oriental topaz for yellow, but these misleading terms are discouraged today. Instead sapphire is prefixed by 'fancy', or the colour, for instance pink sapphire. Only one colour is known by a unique name – the rare padparadscha. This delicate pinkish orange to orangey pink is said to be the romantic combination of the lotus blossom

▼ The highly subjective colour range of padparadscha sapphire. 9.35 ct (left) and 57.26 ct from Sri Lanka (right).

and the sunset in Sri Lanka, the country where it was discovered. The name is derived from the Sinhalese for lotus flower colour.

Large sapphires are rare, although more common than ruby, and can reach hundreds of carats. Most gem-quality faceted sapphires, however, are less than 5 ct and padparadscha is rarely over 2 ct. The Logan Sapphire, housed in the Smithsonian Institution in Washington, DC, USA, is one of the largest faceted blue sapphires in the world. This flawless cushion-cut gem hails from Sri Lanka, most likely the Ratnapura area, and weighs 422.98 ct. Another of the world's most famous sapphires, a 12 ct oval gem set in a halo of diamonds, is in the engagement ring of Her Royal Highness The Princess of Wales, which was inherited from the late Princess of Wales. The largest padparadscha is thought to be the 100.18 ct oval gemstone held in the American Museum of Natural History, New York.

Sapphire occurs in all colours (bar red), colourless, and particoloured. The most valued colours are intense velvety mid-blue known as cornflower blue or Kashmir blue, and deep saturated royal blue. A hint of violet is also desirable, and value increases with colour saturation. Padparadscha's more delicate colour is just as highly valued, with top quality gemstones selling for tens of

thousands of US$ per carat. The colour however is subjective for such an expensive gem, and there is disagreement as to when a padparadscha should be simply termed orange or pink sapphire. Similarly, there is no official boundary between ruby and pink sapphire for light red stones. Colour change sapphire is known, typically ranging from blue or violet in daylight to purple or reddish purple in incandescent light.

As colour is key to value, it influences how the gem is cut. The colour is often zoned, so gemstones are cut with the area of highest colour concentration in the culet, causing the entire stone to appear uniformly coloured when viewed face up. Coloured sapphire is also strongly pleochroic, with blue stones exhibiting violet-blue/greenish blue. Cutters will orientate the rough to obtain the more desirable violet-blue.

Iron is the most common impurity, creating green and yellow. Blue requires both iron and titanium to be present, but only at a concentration of 0.01%. Pink is caused by chromium, and padparadscha is

▲ This mostly colourless 1.92 ct sapphire is cleverly cut with the blue zone in the culet, to appear uniform blue faceup.

◀ Sapphires from Rock Creek, Montana, USA showing a wide colour range, individually weighing up to 4.62 ct.

produced by chromium combined with a defect in the crystal structure. Combinations of impurities create other colours, such as purple from iron and titanium with chromium. Many black and dark brown sapphires are coloured by micro inclusions, and may actually have a different body colour. Colour change sapphires can be caused by the presence of vanadium.

Iron-bearing sapphires (blue, green, yellow) may exhibit a distinctive absorption spectrum with three absorption lines in the blue. Purple, orange and some natural colour change sapphires may show a combined iron and chromium spectrum. Synthetic colour change sapphire can have a distinctive spectrum from vanadium. While few iron-bearing sapphires fluoresce, there are some notable exceptions, such as yellow and orange Sri Lankan stones which show a characteristic apricot-yellow fluorescence. Zoned chalky bluish fluorescence following the crystal growth, visible under shortwave UV light, can indicate heat treatment.

Sapphire tends to have better clarity than ruby, but is rarely flawless. Silk is the most prevalent inclusion, and fine clouds of microscopic particles can scatter the light giving a velvety appearance without affecting transparency. This haziness is sometimes described as sleepy, and is distinctive in blue sapphires from Kashmir. These cornflower blue sapphires, which set the standard for colour, were discovered

▶ 1.10 ct particoloured Australian sapphire, heat treated to lighten the colour, cut by John Dyer Gems.

around 1881, at almost 4,500 m (15,000 ft) above sea level. It is said a landslide revealed the blue crystals. Many sapphires were mined until 1887 when the deposit was considered exhausted, with only sporadic production since. These rare and magnificent gemstones occasionally appear in older jewellery, and command high prices.

Star sapphire occurs in most colours with blue highly desired. Black or dark brown star stones are common, caused by hematite-ilmenite inclusions. Star sapphire is fashioned en cabochon to reveal the asterism, however black stars are more fragile due to the numerous inclusions, and partings along the planes in which they lie, so are cut with shallower domes.

Some sapphires exhibit a trapiche-like arrangement of inclusions, in a pattern resembling six spokes of a wheel. However, as these are not growth sectors, it is not a true trapiche. These are known from many locations including Vietnam and Australia.

Sapphire has both metamorphic and magmatic origins. Those formed through metamorphism occur in marble, gneiss and schist, in a range of colours including padparadscha. Magmatic sapphire is associated with alkali basalts, and is high in iron, typically forming blue, green and yellow suites, with fancy colours less common. The parent rock and geological origin greatly affect the value, directly influencing clarity, colour, and even carat weight and cutting style which are constrained by the shape of the crystal.

▲ Blue Kashmir sapphire with a 'sleepy' appearance caused by minute silk inclusions.

Myanmar has produced fine sapphires for centuries, famous for their royal blue colour. Star sapphires and fancy colours are also found. The sapphires originate from syenites associated with the ruby-bearing marbles, and are mined in smaller quantities from alluvial gem gravel deposits. Sri Lanka has produced sapphires for over two millennia and is known for the largest gemstones. It remains an important source for blue and star sapphire. These have lower iron content and are light to medium blue with a hint of violet. Sri Lanka also produces a variety of fancy colours and is the main producer of padparadscha. Large quantities of milky white geuda are mined which respond well to heat treatment, increasing the amount of affordable sapphire on the market. Sapphire is mined from extensive gem gravel deposits in the Ratnapura and Elahera areas. The origin is metamorphic but very few primary deposits have been found. Thailand and Cambodia have produced sapphires from significant deposits located along their border in the Chanthaburi and Palin areas, mined from placer deposits. The sapphires are associated with basalts, reflected by high iron imparting a dark inky blue with hints of green. Yellow and green sapphire, as well as golden and black star sapphire, are also found. Vietnam has several deposits associated with metamorphic hosts in the north, producing grey, blue, violet and padparadscha. Basalt-associated blue, green and yellow sapphires are produced in the south. All are mined from alluvial placer deposits.

Several major finds have led to Madagascar becoming one of the most important suppliers today, thought to be producing more fine blue sapphires than Sri Lanka and Myanmar combined. In 1994 extraordinary sapphires of metamorphic origin, with colour to rival Kashmir, were found at Andranondambo in the south. Soon afterwards blue, green and yellow basalt-associated sapphires were discovered in the north, then in 1998 large placer deposits of blue and fancy colours were found at Ilakaka, in the southwest. Madagascar has an incredible wealth of coloured stones, and most mining is done by hand by artisanal miners. East African countries produce a range of sapphires, particularly Tanzania (including pastels and dark orange classified by some as padparadscha), Kenya (including particoloured), Mozambique, Malawi, Zimbabwe (including stars) and Ethiopia. Nigeria and neighbouring Cameroon are also big sources with inky blue, green, yellow and particoloured sapphires associated with basalts.

Australia is an important source, commercially mined for the past century from placer deposits in New South Wales and Queensland. The sapphires are associated with alkali basalts, and have high iron content and dark colour. They occur dominantly in inky blue, green, yellow and particoloured, and less commonly in fancy orange, pink, purple and star stones. Production has declined in the past few decades. Sapphires are also found in Tasmania, Victoria and Western Australia.

Montana in the USA contains several sapphire deposits with unusual magmatic sources unrelated to the three formation periods. The unique Yogo Gulch is famed for its high clarity and uniform blue to violet colour that does not require heat treatment. Sapphires are mined from a lamprophyre dyke, first discovered in 1895. Faceted gemstones are small due to the thin tabular crystals, considered rare over 1 ct. Production is small and sporadic, but demand within the USA is high. Other deposits produce greyish blues and a range of fancy colours, mined from alluvial gravels. Other minor sources include China, Brazil, Pakistan, Colombia, Russia and Scotland.

The majority of sapphires on the market are thought to be heat treated. Untreated sapphire attracts a premium of at least 50%, so disclosure and detection are crucial. Heating to 800–1,200°C will lighten the dark blues of basaltic sapphires. Heating to 1,200–1,800°C dissolves silk allowing the titanium to be absorbed into the crystal structure. This improves transparency, and also colour in

▲ Faceted gemstones up to 4.62 ct (left) and rough sapphire crystals (right) from Rock Creek, Montana, USA.

iron-bearing sapphires, by providing titanium to create blue. Milky geuda from Sri Lanka has little colour but lots of silk, and heating has a dramatic effect producing high clarity blue, orange or yellow stones. High temperature and pressure treatment is now becoming common, used to enhance colour. Heated sapphires may show a chalky blue fluorescence under shortwave ultraviolet light. The resultant colour from heat treatment is stable.

Sapphires are treated by diffusion to enhance or alter the colour. Gemstones are heated to very high temperatures (1,700–1,900°C) in the presence of certain elements that diffuse into the crystal structure. Surface diffusion of titanium into colourless or pale sapphires creates saturated blues, used since the late 1970s. This forms a shallow layer and does not penetrate the whole stone. Titanium can also enhance silk, producing asterism. Around 2001, large quantities of padparadscha sapphire appeared on the market, sparking suspicion. Gem laboratories determined a new treatment diffusing beryllium into pink sapphire, imparting a yellow colour to give the appearance of padparadscha. As beryllium is a smaller atom than titanium, the diffusion can penetrate the whole stone, called lattice diffusion. This treatment produces bright yellow and orange, and will lighten dark blue.

Diffusion treatment requires disclosure and may sometimes be detected by immersing the stone to reveal colour concentrations following the shape of the faceted gem not the crystal growth, along with inclusions damaged by the high heat. Some lattice-diffused sapphires, however, require advanced analysis by gem laboratories in order to detect the treatment.

Rarely, colourless or pale sapphires are irradiated to create bright orange padparadscha colours, which are not stable and fade with exposure to light. Synthetic sapphire is created in all colours by the addition of different impurities. It is predominantly grown by flame-fusion, and less commonly by flux melt, hydrothermal and crystal-pulling (Czochralski) methods. Flame-fusion stones generally do not fluoresce or show the full distinctive iron absorption spectrum. Vanadium is used to create a synthetic colour change sapphire with a purplish pink to slate-blue colour change. This is used as a simulant of alexandrite, although the colour more resembles the pleochroism of amethyst. The vanadium produces a diagnostic absorption spectrum.

Sapphire is imitated by variously coloured glass and plastics. Doublets are created with a natural green sapphire slice crowning a synthetic blue pavilion, or a slice of garnet on cobalt blue glass. The gemstone appears blue with natural inclusions when viewed face up. This may be detected by bubbles trapped in the join, and garnet will give a different refractive index.

Many other gemstones can be confused with sapphire. Blue tanzanite, kyanite, spinel, benitoite, and tourmaline have a similar appearance, and iolite even has the misleading trade name water sapphire. Fancy sapphires may be confused with chrysoberyl, topaz, citrine, amethyst, kunzite, beryl varieties, spinel and tourmaline. Colour change spinel exhibits a similar colour change of blue to purple-pink. These can all be distinguished by different optical properties.

Diamond

Diamond is the quintessential gem. A unique combination of incredible lustre, brilliance and fire, makes it exceptionally valuable and desirable, outselling every other gemstone. Well suited for jewellery, it is the hardest known natural substance, scratched only by another diamond. Its name is derived from the Greek 'adamas' for unconquerable or invincible stone. Diamond is treasured not only for its beauty and properties, but also its science – forming deep underground millions of years ago, it is a time capsule bringing information about inner Earth to the surface.

Diamond

Composition C carbon **Crystal system** isometric **Hardness** 10 **Cleavage** perfect **Fracture** conchoidal **Lustre** adamantine **SG** 3.50–3.53 **RI** 2.415–2.420 **Birefringence** none **Optical nature** isotropic **Dispersion** 0.044

Diamond is composed of a single element – carbon, and has the same chemical composition as graphite, but the carbon atoms are arranged and bonded in different orientations, giving them vastly different properties. Graphite is opaque black and soft, while diamond is transparent, much denser and incredibly hard. So how does carbon form diamond instead of graphite? Diamond crystallizes under high pressure and high temperature conditions. It requires a minimum of 900 to 1,250°C (1,650 to 2,280°F) and 4 to 5 GPa of pressure. These conditions are reached at a depth of at least 80 km (50 miles) below Earth's surface. At shallower depths, the carbon crystallizes into graphite, the stable low pressure and low temperature form of carbon. The pressure required for diamond has been described as equivalent to the Eiffel Tower inverted on a soft drink can. Diamond generally forms at depths of 120 to 200 km (75 to 124 miles) in Earth's upper mantle. Some diamonds, known as superdeep diamonds, form

in the mantle below the lithosphere (the crust and the uppermost rigid portion of the upper mantle) at depths of 400 to 750 km (249 to 466 miles).

The source of the carbon is speculated to be carbon-bearing fluids that react with the ultrabasic rocks (peridotite and eclogite) of the upper mantle. This releases the carbon, allowing it to crystallize into diamond. Later, the diamond is transported to the surface in molten rock (magma) in rapid, violent volcanic eruptions. The fast ascension and quick cooling of the magma prevents the diamond converting to graphite.

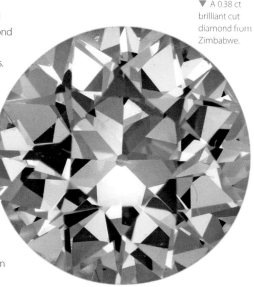

▼ A 0.38 ct brilliant cut diamond from Zimbabwe.

◀ 'The Dresden Green', the world's most famous green diamond. The naturally coloured, 41 ct pear-shaped gem is set in a magnificent hat ornament.

When diamond is mined from its host kimberlite or lamproite pipes, it is known as a primary deposit, and is extracted by large-scale, mostly open pit, mines. The quality and size of the diamond, plus the location and accessibility of the deposit, greatly influence the viability of the mine. Diamond grades range from 1 ct per 100 tonnes (110 tons) where the diamond is of high gem value, to 500 ct per 100 tonnes of predominantly industrial-quality diamond.

Kimberlite and lamproite are easily weathered and eroded, especially after millions of years, so diamond is rarely found in situ on the surface. Due to its high hardness and density, it is often transported far from its source, accumulating with other heavy minerals in secondary placer deposits. Higher quality diamond travels the furthest, as heavily included or fractured crystals are broken up along the way. Alluvial diamond deposits are found in many countries, mined from streams and rivers. Marine deposits are found along the coast of Namibia and South Africa, where diamond is extracted from the beach and seafloor. Although challenging to mine, these deposits are economical due to the higher percentage of gem-quality diamond. Many secondary diamond deposits can be traced back to their primary surface host rock, but the provenance of some is still a mystery.

These special eruptions create a conduit to the surface with a funnel or bowl-shaped top, known as a pipe. There are two different types of eruptions: one forms a rock called kimberlite, and the other a similar rock called lamproite. Both are mixtures of the crystallized magma and fragments carried up from the upper mantle. Kimberlite pipes are the main source of diamond.

Not all kimberlite pipes contain diamond, however, only those that have intruded into the oldest, thickest parts of Earth's crust (which have been stable for more than 2.5 billion years) are diamondiferous. The kimberlite intrusions are much younger, with most less than 550 million years, but ranging from 1.2 billion to 50 million years old. They relate to large-scale geological events such as the breakup of the supercontinent Gondwana, or younger gigantic volcanic events. Diamond is much older than its kimberlite host. The oldest known diamonds are 3.5 billion years old, while the youngest are 660 million years old. The age of a diamond is determined by the age of its inclusions – the tiny solid minerals thought to form simultaneously with their host, which are then captured and preserved inside it. Interestingly, a diamond's age cannot be determined by radiocarbon dating methods, as the radioactivity of carbon-14 fades in tens of thousands of years, and diamonds are far older.

Other methods of formation include high pressure metamorphism, and the impact of extra-terrestrial material creating high temperature and high pressure conditions. The diamond crystals formed however are small, generally less than 0.5 mm (0.04 in), and of no significance as gemstones.

The first discovery of gemstones that can be reliably interpreted as diamond dates from 300 BC in India, and India was the only known source of diamond for more than 2,000 years. It was mined from alluvial placer deposits, producing large colourless stones, as well as a range of colours.

The Golconda mines are the most well known and are said to be the source of many legendary diamonds including the Koh-I-Noor (now in the British Crown Jewels) and the Dresden Green. The production from India was small but steady, and was reserved for the wealthy, including royalty. As Indian supplies declined, the discovery of alluvial diamond deposits in Brazil provided a second source. Found around 1725 by gold miners, Brazil supplied most of the world's diamonds for over 100 years and small quantities are still extracted today. The deposits are predominantly alluvial and the source pipes for many have not been found. Brazil has produced a number of significant stones including pink diamonds, and the largest diamond ever found - a 3,167 ct carbonado. Carbonado is a unique type of diamond, very different to other diamond, found only in Brazil and the Central African Republic. It is an opaque black or brown porous aggregate of tiny interlocking crystals, and this polycrystalline nature gives it higher toughness. Due to its excellent abrasive properties, carbonado was popular in industry before the invention of synthetic diamond. Its host rock and its formation process are still a mystery, with theories including extra-terrestrial origin.

The discovery of diamond in South Africa occurred in 1866, with a yellow 21.25 ct crystal found on the banks of the Orange River, now faceted into the 10.73 ct Eureka diamond. In 1870 diamond was found in situ in its host rock for the first time, and the host was named kimberlite after the city Kimberley. These 'dry diggings' yielded a much higher concentration of diamond than alluvial deposits, and large quantities became available. This included the first significant find of yellow diamond, found in the Cape Province, from which the name Cape diamond originates. By 1889 South Africa

was supplying 99% of the world's diamonds, continuing at this rate for several decades. The Kimberley and Dutoitspan mines were both significant early sources, and the Premier Mine, now Cullinan, was historically a producer of rare large crystals, including the largest gem-quality diamond found to date, the famous Cullinan Diamond. Discovered in 1905, it weighed 3,106.75 ct (621.35 g, 1.37 lbs) and amazingly was only a cleaved portion of a larger crystal, the rest sadly never found. It was cut into nine major and nearly 100 minor gemstones. The two largest are the Cullinan I, which at 530.2 ct remains the largest colourless faceted diamond in the world, and the Cullinan II, weighing 317.4 ct. Both are in the British Crown Jewels.

Namibia's first diamonds were found near Lüderitz in 1908, leading to the discovery of huge placer deposits along the coast. Here the beaches are mined, and boats off the coast suck up sediment from the sea floor. Aeolian deposits are also mined, accumulated by winds in coastal desert sand dunes.

▼ A complex crystal weighing 0.95 ct, possibly the first confirmed diamond found in Australia in 1851 from Ophir, New South Wales.

The earliest recognised diamonds in Australia were recorded in 1851 from the Bathurst area, New South Wales, predating South Africa's finds. Australia is best known for the Argyle Mine in the north of Western Australia, and it was the first place diamond was found in a lamproite pipe. Argyle produced a high proportion of gem-quality coloured diamonds, of which nearly 80% were brown. Argyle is famous for pink diamond and supplied around 90% of the world's market, yet these rare pinks were only 1% of their gem-quality production. The pinks are small, with the largest rough diamond found weighing 12.76 ct. Even rarer are blue, grey and violet colours. Argyle began production in 1983, reaching 40% of the world's production by carat in 1993. Sadly it has reached the end of its commercial life, closing at the end of 2020.

Russia, Botswana and Canada are major sources today by both carat weight and value. Russia has been an important source since the 1960s, today providing nearly 30% of global production by carat and value, producing a range of colours including pinks. Botswana began mining in 1970 and today produces more than 15% by carat, but nearly 25% by value, of the world's production. It contains some of the largest, and the highest value, mines. Canada began mining in 1998 with important deposits opened in the following decades. The mines are in the north and, for some, mining is restricted to certain times of the year due to the extreme weather conditions. The world's oldest known diamonds, approximately 3.5-3.3 billion years old, were found at the Ekati and Diavik Mines in Canada.

The Letšeng diamond mine in Lesotho, in southern Africa, has produced the highest percentage of large crystals in recent years, with the biggest to date weighing 910 ct. It has a low production but, due to the large diamonds, has the highest per carat value of any diamond mine. Other sources include Angola, Zimbabwe, Namibia, Ghana, Sierra Leone, Democratic Republic of the Congo, and the USA. Global production has increased from around three million carats at the start of the 1900s, peaking at 176 million carats in 2006, and settling around 142 million carats in 2019.

The ethical, political, economic and environmental impact of diamond has been under the spotlight in the past few decades. Consumers wish to know the geographical origin of their diamond, to be confident it was mined responsibly and sustainably. Conflict-free diamonds are those that do not fund terrorism or armed conflict (those that do are known as conflict diamonds), and also includes factors such as fair labour and environmentally responsible mining. Several countries, such as Botswana and Canada, are popular for their ethical diamonds. The diamond industry has had a measurable positive impact on their economy and local communities, providing employment and protecting the environment

▼ Pear-shaped fancy colour diamonds (left to right) 0.63 ct pinkish orange, 0.43 ct intense pinkish purple, 0.40 ct vivid purplish pink.

through conservation or strict laws. In 2002, the United Nations and major diamond-producing countries created an international certification scheme to regulate the trade of rough diamonds. Known as the Kimberley Process, it aims to prevent conflict diamonds being sold on the gem market. Since 2003 all rough, non-faceted diamonds (including sawn and cleaved) must receive a Kimberley Process certificate in order to cross any international border, to certify the rough was not sold to fund a conflict. Some countries are not currently participating or are banned from international trade. It is critical to keep up to date with the Kimberley Process to ensure that any import or export of uncut diamonds is legal (www.KimberleyProcess.com).

◀ A 3.26 ct octahedral crystal of diamond.

▶ A 0.39 ct distorted dodecahedron with curved crystal faces from Oven's Goldfields, Victoria, Australia.

Pure diamond, containing only carbon, is colourless. However, most contain impurities substituting for carbon atoms. While only minute quantities may be present, measured in parts per million, this is enough to impart different colours and fluorescence. Scientifically, diamond is defined by these impurities and classed by the presence or absence of the most common one, nitrogen:

Type I contain impurities of nitrogen, and 98% of all natural diamonds are this type. Type I diamonds are subdivided:

Type Ia have clusters of nitrogen atoms in various arrangements, and a nitrogen content up to 3,000 ppm. The majority are known as Cape diamonds, which range from near colourless to yellow, often with a hint of brown. Many fluoresce a characteristic blue under longwave UV light. Type Ia may also be colourless, and rarely pink, blue, violet or chameleon. The majority of Type I diamonds are Type Ia.

Type Ib have nitrogen dispersed as single atoms. The nitrogen content is around 100 ppm. These diamonds often have a saturated golden yellow colour described as canary and occasionally exhibit a weak orangey-yellow fluorescence. Type Ib are rare.

Type II lack any detectable nitrogen. Superdeep diamonds are often this type. Type II diamonds are subdivided:

Type IIa are the purest, usually colourless, but may be pale pink, yellow or brown. They generally do not fluoresce. Many of the large colourless, high-clarity, famous diamonds such as the Cullinan and Koh-I-Noor are Type IIa.

Type IIb have impurities of boron, typically 0.1 to 2ppm, giving a blue colour, with a blue-green or red phosphorescence. These are the only type to conduct electricity. The most famous example is the Hope Diamond.

Diamond has isometric crystal symmetry. The carbon atoms are strongly bonded in four directions, which gives it incredible hardness.

▲ A diamond macle with tabular triangular shape, and trigons on the surface. 1.4 cm (0.5 in) height.

▲ Triangular etch pits known as trigons are common surface features on diamonds. They are caused by the resorption process.

▼ A red garnet inclusion is retained as a feature in this 1.98 ct step cut diamond.

Diamond crystallizes in two primary crystal forms: octahedrons and less frequently cubes. Other forms include macles (spinel-law twinned octahedral crystals with a tabular triangular shape), polycrystalline aggregates, fibrous, and complicated rounded shapes. The rounded shapes are secondary forms due to partial resorption of the original crystal, dissolving through reaction with the kimberlite magma as it is brought to the surface. It is thought up to half of the original crystal may be resorbed. They have curved surfaces resembling shapes such as dodecahedrons and tetrahexahedrons.

Of all diamond mined, only around 20% by weight are deemed transparent gem quality. For the jewellery and gem market, inclusions are normally considered detrimental to the diamond's quality, but for scientists they are incredibly important. Crystals of other minerals and fluids are encapsulated in the diamond, deep within Earth, and preserved for millions of years. Studies of these inclusions provide a glimpse into the history of our inner planet, providing clues about the depth and geological environment in which the diamond grew. Minerals such as olivine, chrome diopside and chrome-rich pyrope garnet indicate peridotite host rocks, and almandine garnet may indicate eclogite. Other

common inclusions are growth features such as zoning and twinning lamellae. The high-stress conditions of growth and turbulent transport to the surface can cause internal fissures, insipient cleavage, or graining – a haziness or phantom lamellae seen inside the diamond due to deformities of the crystal structure on an atomic scale.

Its unique combination of properties makes diamond highly desirable as a gem. The high refractive index gives an incredibly bright, mirror-like lustre, called adamantine (named for diamond). The refractive index also gives diamond high brilliance, creating the bright sparkle. Diamond's dispersion of 0.044, surprisingly classed as moderate, is higher than most other well-known gems, and creates the rainbow flashes of fire for which it is so well known.

Diamond is the hardest known natural substance, honoured as 10 on the Mohs scale, and allowing gem cutters to achieve the highest quality of cut, with sharp facet edges and mirror-flat faces. It also makes diamond ideal for every type of jewellery. However, what few consumers realise is that diamond has a perfect cleavage, so while it cannot be scratched, it is fragile and can cleave if knocked or dropped. The reason that one diamond can cut and polish another is because the hardness varies slightly in different directions across the crystal structure. Diamond grit or powder used on polishing wheels consists of particles in all different orientations, meaning that some will always be orientated in the hardest direction allowing them to cut another diamond.

These bright, sparkling, fiery gems have captured the hearts of many, and have long been used in jewellery. The earliest known, occasionally found from the first few centuries AD, are natural octahedral crystals set into Roman rings, providing evidence to support trade routes from India. Philosopher Pliny the Elder refers to 'adamas' in his book *Historia Naturalis* (77 AD), which is interpreted by many historians to be diamond. Diamond made its way into European jewellery in the 1300s and the modern diamond market began with the discovery of this gem in the late 1800s in South Africa. Successful marketing campaigns by De Beers in the 1940s made diamond the premier choice for engagement rings, and today around 80% still are diamond.

Faceted diamonds are graded by the 4Cs – cut, carat, clarity and colour. This system was developed by the Gemological Institute of America (GIA) and is considered by most to be the industry standard, allowing consistency across the international market. Gem laboratories assess diamond gemstones by these, or similar, ratings.

Cut

The history of fashioning diamond has changed dramatically, from measured by eye and cut by hand to the precise and standardized techniques of today. The earliest polished diamond gemstones were simply octahedral crystals with polished faces. The discovery that diamond could be cleaved into smaller shapes, followed by the invention of the polishing wheel and bruting machine, allowed the creation of symmetrical facets at specific angles. Today precise laser cutters, virtual 3D modelling and other equipment allow the best yield from the rough with fast, accurate and symmetrical fashioning.

The shape and quality of cut have huge impact on the appearance of a diamond as cut determines the way the gem interacts with light. It affects the dispersion (fire), brilliance (light return), scintillation (sparkle as the gem is moved) and can also influence the perceived colour. Cutting is always a compromise between retaining carat weight and creating the perfect sparkling gemstone, to obtain maximum value.

▼ Brilliant cut diamond showing brilliance and fire, with contrasting light and dark internal areas.

▼ The angles and proportions of a brilliant cut influence the cut quality, affecting its brilliance, scintillation and fire

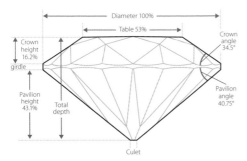

- Diameter 100%
- Table 53%
- Crown height 16.2%
- girdle
- Crown angle 34.5°
- Pavilion height 43.1%
- Total depth
- Pavilion angle 40.75°
- Culet

The size and shape of the finished gem is influenced by the overall shape of the rough diamond, plus any inclusions and colour zoning. Years may be spent planning the cutting of a rare, extremely large rough to ensure a spectacular gem with highest yield. The quality of the cut is determined by the symmetry, proportions and polish. The overall shape should be symmetrical, with facets even and aligned.

There are several cuts typically used for diamond. Faceted styles for colourless diamond, such as the brilliant cut, are specifically designed to refract and reflect as much light as possible back towards the viewer to achieve maximum fire and brilliance. Other styles, such as mixed cuts, are used to emphasize colour over fire and brilliance. For fancy colour diamond this intensifies the colour, improving the grade and increasing the value. These styles, however, are detrimental to near colourless gemstones making them look less white.

The brilliant cut is the most prevalent cutting style, designed in 1919 by Marcel Tolkowsky. The modern round brilliant cut has 57 facets (or 58 with a facet on the culet) creating bright, lively gemstones with exceptional sparkle. By using the refractive index of diamond, the exact angles of the facets and ideal proportions of the crown and pavilion can be determined, so that all light entering the diamond through the crown is reflected back towards the viewer. This is known as total internal reflection, as no light leaks through the back of the gemstone. Brilliant cuts are fashioned in a range of outlines including heart, pear, oval, marquise, cushion, and round (the most popular). Gem laboratories evaluate the quality of round brilliant cuts, measuring how close the proportions are to ideal, evaluating the size of table, height and angles of the crown and pavilion, and the girdle and culet. The best stones are bright with an even pattern of contrasting light and dark areas seen inside the stone.

▼ Brilliant cut diamonds of differing cut quality. Excellent with total internal reflection (left) to poor with light leaking out the back, creating dark areas (right).

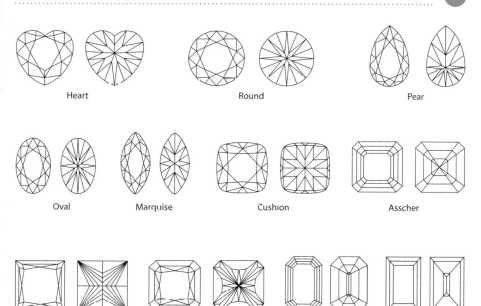

Heart	Round	Pear
Oval	Marquise	Cushion	Asscher
Princess	Radiant	Emerald	Baguette

▲ Illustrations of the different cutting styles used for diamond.

Step cuts in the shape of a rectangle (baguette and emerald cut) and square (asscher cut) create graceful gemstones with clean lines of reflection. However they do not have the same fire or scintillation as other cuts, nor do they hide inclusions as well. The Princess cut is a trade name for a square modified brilliant cut with chevron-shaped facets on the back. It has high brilliance and a modern look. The sharp corners, however, require protection by the setting. The radiant cut is a modified brilliant cut with a square or rectangular outline and truncated corners. It has more facets on the pavilion to accentuate the colour rather than fire. Heavily included black diamonds are cut to show the high lustre rather than clarity, and often have small tables with many facets on the crown.

Carat

The weight of a diamond is measured in carats and should be stated to at least two decimal places, and is only rounded up if the third decimal is a 9, for example 0.998 = 0.99 ct but 0.999 = 1.00 ct. For a diamond below 1.00 ct, its weight is often expressed as points, where one point is one hundredth of a carat, for example 0.30 ct is 30 points. Melee are small gemstones of less than 0.20 ct, normally round and up to 4 mm (0.16 in) in diameter. They are used as accent stones to give jewellery extra sparkle, for instance in a halo around a central stone.

The final weight of a cut diamond is always a trade-off between maximizing beauty and retaining highest yield. Diamonds of similar cut, colour and clarity are priced based on their weight, and prices are benchmarked across the diamond industry through the Rapaport Price List. Jumps in value occur at set weights such as 0.50 ct and 1.00 ct so a 1.01 ct diamond may be significantly more expensive than a 0.99 ct gemstone. Larger diamonds are valued at more per carat due to their rarity. Diamonds with particularly high weights have been found in recent years, such as the 1,109 ct Lesedi La Rona, found at the Karowe mine in Botswana in 2015, and the second largest colourless gem-quality diamond to date. The discovery of these 'oversized' stones has led to mines increasing the size settings on their automatic sorting machines to ensure the larger rough are not considered chunks of rock and accidentally sent to the crusher.

▲ 0.82 ct radiant cut fancy intense purplish pink diamond with visible inclusions.

Clarity

A diamond's clarity is judged by the presence (or absence) of flaws, as internal inclusions or external surface imperfections. A diamond with top clarity is flawless. Diamond is graded by assessing the number of inclusions, their size, location, colour and relief – how much they stand out. The position of an inclusion affects the grade, depending on whether it is central or towards the edge of the gemstone. The inclusion's colour affects how visible it is. When viewing a faceted diamond face up there is an internal pattern of light and dark areas, which gives diamond its sparkle. Dark inclusions in areas that remain dark as the gem is moved are hidden compared with dark inclusions in areas that are light, and vice versa for light-coloured inclusions. The more included a gemstone is, the less fire and scintillation it has.

The GIA clarity grading system uses flawless (FL) as the top grade where there are no visible internal inclusions or external blemishes when examined under 10x magnification. The other categories are internally flawless (IF), very very slightly included (VVS$_1$ and VVS$_2$), very slightly included (VS$_1$ and VS$_2$), slightly included (SI$_1$ and SI$_2$), included (I$_1$, I$_2$, I$_3$). Other grading systems may be based on this with additional divisions. Flawless diamonds are rare, and most diamonds are in the SI to I clarity grades.

Colour

Diamond occurs in a variety of colours. Gem-quality diamond is predominantly pale yellow to pale brown. Colourless is quite rare, therefore the closer a diamond is to colourless, the higher the value. Near colourless diamonds are graded by their degree of colourlessness and the slightest difference – unnoticed by the untrained eye – can affect value. The GIA's grading system is the most widely accepted using an alphabetical scale D to Z. D is the top grade for colourless (D–F), then near colourless (G–J), faint (K–M), very light (N–R) and light (S–Z), with Z a light yellow or brown tint. Stones with lower colour grades will show less brilliance and sparkle. The reason the grading does not start at A is to be distinct from other grading systems, as D is not normally associated with the highest quality. Colour grading is undertaken by comparing gemstones to a set of stones of known colour under controlled lighting and viewing conditions.

One in 10,000 faceted diamonds is termed fancy colour. This includes any yellow and brown diamond outside the D to Z range, and all other colours. Yellow and brown are the most common fancy colours, and blue, purple, pink, red and green the rarest and most desirable. Subtle

D   H   N   Z

◀ GIA colour grading system uses D (colourless) to Z (light colour).

changes in the colour can greatly impact the value. Fancy colour diamond is described by its hue (colour), tone (lightness or darkness) and saturation (strength of colour). Grading is based on the intensity of colour with grades fancy light, fancy, fancy intense, fancy vivid, fancy dark or fancy deep. Coloured diamond grading is more subjective than near colourless, and systems vary between different gemmological laboratories.

A high proportion of fancy colour diamond is yellow, found from many locations around the world. The majority are Cape diamonds (Type Ia) with colours caused by clusters of nitrogen impurities. These are yellow, from Z grade to darker and may have brownish or greyish modifying colours. The most desired is the pure bright yellow termed fancy vivid yellow. Much rarer, and considered by some to be more beautiful, is the intense canary yellow to a hint of orange (Type Ib), caused by single isolated nitrogen atoms replacing carbon atoms.

Blue diamond is extremely rare, tending towards greyish blue. Most blues contain impurities of boron substituting for carbon atoms (Type IIb), and the higher the boron content, the richer the blue. Blue diamond is most well known from the Cullinan Mine in South Africa, and historical Indian mines. In 2015 the 12.03 ct Blue Moon of Josephine diamond, from the Cullinan Mine, sold at auction for more than US$4 million per carat (£2.85 million) – the highest ever price per carat of any gemstone.

Blue, grey and violet diamonds from Australia's Argyle Mine contain impurities of hydrogen and nitrogen, thought to be associated with the cause of colour. Some diamonds from there show a colour-change effect from violet in incandescent light to blue in daylight.

▶ 1.47 ct pear shaped Cape diamond of pale yellow hue.

▼ 0.67 ct brilliant cut canary yellow diamond with a hint of orange.

▶ The famous Hope Diamond, 45.52 ct, coloured blue by traces of boron impurities averaging 0.5 ppm.

▼ 0.52 ct fancy deep grayish violet heart.

▲ 1.13 ct fancy deep orangey pink heart.

▼ 0.40 ct pear shaped fancy vivid purplish pink.

Pink diamond is very rare and highly desirable. Pale to vivid in colour, it also ranges into purplish, brownish or reddish hues. Most are small, up to 1 ct, and incredibly rare over 10 ct. The Argyle Mine was the main source, which produced 90% of the world's pinks, however its closure will dramatically reduce availability. Pinks have also been found in India, Brazil, South Africa, Canada and Russia. Almost all – 99.5% – of pink diamonds, as well as rare purples and even rarer reds, show graining (parallel lamellae of colour concentrations). These diamonds are Type Ia or Type IIa. The exact cause of colour is still unknown but is associated with the post-growth distortion of the crystal structure (known as plastic deformation). The deformation is thought to be caused by the stress of the high-pressure, high-temperature conditions in which the diamond formed. Interestingly, this more complicated crystal structure means these pinks take longer to polish than colourless diamond. The extremely rare other 0.5% of pink diamond are Type IIa of uniform pale pink with no graining, known as Golconda pink. These show a yellow-orange-red fluorescence. Top-quality pink diamond of vivid colour can sell for more than US$2.5 million per carat.

Brown is one of the most common diamond colours, and therefore a common fancy colour. Up until the 1980s, brown diamond had little value on the gem market and was used for industrial purposes. Following the opening of the Argyle Mine, which produced high quantities of browns, successful marketing strategies and tasteful new names such as champagne, cognac and coffee, increased their popularity. The cause of colour for most brown diamond is associated with plastic deformation of the crystal structure, causing defects of clusters of missing carbon atoms, called vacancies. This is similar to the cause of pink colour, and they likewise show colour graining. Brown diamond can be treated by high pressure high temperature (HPHT) treatment to decolourize or improve the colour.

Green diamond is one of the rarest natural colours and occurs through radiation damage of the crystal structure creating vacancies of missing carbon atoms. This may occur naturally underground due to nearby radioactive minerals. The colour is often concentrated at the surface of the diamond crystal, requiring careful cutting so as not to remove it through faceting. Diamond may be artificially irradiated producing a darker blue-green colour.

Chameleon diamond is a curious gem that temporarily alters its colour. It typically changes from yellowish to brownish green after exposure to light for a minute or so. The colour change can be reversed when the diamond is stored in the dark for several days or gently warmed.

Naturally black diamond is uncommon, and most on the market are treated. The majority of natural black diamonds are coloured by nano- to micro-sized mineral inclusions such as graphite, pyrite or hematite. Additionally, they may contain fractures, radiation stains, or be polycrystalline. Most are opaque, but translucency varies, and the diamond can actually be near colourless, but is so heavily included it appears black. Treatments to produce black diamond include heating heavily fractured or included stones to induce graphite formation. Heavily irradiating a diamond creates a dark green that appears black when faceted. Shining a strong light through these gemstones reveals the transparent green colour. Black diamond has become popular in the past 30 years, particularly when cut to utilize the adamantine lustre. Inclusions and fractures make black diamond challenging to cut and polish, and it requires greater care as it is more vulnerable to damage.

Fancy white is a rare and unusual diamond with a translucent milky white appearance that may be opalescent. This is predominantly caused by numerous tiny inclusions, which scatter the light. Similar to fancy black, fancy white diamond ranges in transparency.

Both coloured and colourless diamonds can fluoresce under longwave ultraviolet light. They show a range of colours, most frequently blue, then

▲ 1.37 ct heart shaped fancy deep yellowish orange.

► 1.07 ct radiant cut fancy intense green diamond of natural colour.

◄ 6.15ct chameleon diamond with reversible colour from yellow (left) when stored in the dark to olive green (right) after exposure to light.

◀ 1.03 ct heart of opalescent fancy white colour.

▶ 0.34 ct cushion cut fancy vivid green-yellow (top) exhibiting strong green fluorescence under longwave UV light (bottom).

yellow, and more rarely colours such as orange, white, and green. Fluorescence is described as none, faint, medium, strong or very strong. Just over one third of colourless to near colourless diamond fluoresces, with almost all (97%, all Type Ia) having a blue response. Blue fluorescence is often considered detrimental to the overall appearance of the stone, particularly in colourless gems, and those with medium to strong fluorescence have lower values. In a tiny portion of diamonds, very strong blue fluorescence can create a hazy or oily appearance. However, in general, little difference can be seen in daylight between faceted stones with and without blue fluorescence, and it can even have a desirable effect making slightly yellowish diamond appear more colourless. It is important to note that fluorescence may affect the colour grading of a diamond depending on the light source used, so when purchasing be sure to view the diamond in different types of lighting.

Blue diamond produce a blue-green to whiteish, or red phosphorescence following exposure to ultraviolet light. The Hope Diamond is famed for its characteristic fiery red phosphorescence, which lasts up to a minute after the UV light source is removed.

Diamond is perfect to wear in all jewellery. As well as having incredible hardness, it is chemically inert and resistant to acids, so can be worn every day. Due to its hardness, however, diamond jewellery should not be stored loose as it will scratch other gem materials. Avoid hard knocks, and thermal shock from sudden changes in temperature, which may fracture or cleave the gem. Diamond can also be burned, so care must be taken with a jeweller's torch when setting. Warm soapy water is safe to use when cleaning diamonds.

▼ Diamonds mounted in a butterfly shaped brooch.

There are several methods to enhance or alter the colour of diamond, and all treatments should be disclosed. These include coating with a micro-thin surface layer, which can be detected under magnification by damage to this softer coating. Heavily included and fractured grey or brown diamonds may undergo high temperature low pressure treatment, which causes graphite to form along surface-reaching fractures, turning them black. Irradiation is used to create fancy blue to green colours, and when followed by heating (annealing) can produce pink, red and yellow. Heavy irradiation can produce very dark green which, when cut, gives the appearance of fancy black.

HPHT treatment is used on specific types of diamond with different results. It is commonly used on brown diamond to decolourize, or alter to yellow, yellowish-green, green, blue or pink colours by removing brownish hues. As this treatment is only effective on a small percentage of natural diamond, it is more commonly used on synthetic stones. Chemical Vapour Deposition (CVD) synthetic diamond is often treated in this way, which both strengthens the diamond and reduces brownish colours. Diamond may be treated by multiple processes such as HPHT and irradiation to achieve the desired colour; both treatments are permanent, and difficult to detect, requiring testing in gemmological laboratories.

Diamond clarity can be improved. Surface-reaching fractures may be filled with lead glass, which has a similar refractive index to diamond, reducing the appearance of the fracture. The filling may be detected by a pink or green flash, seen as the stone is rotated, where light reflects off the fracture interface. This treatment is not permanent and can be damaged by heat from a jeweller's torch. Laser drilling has been used since the late 1960s to reduce the appearance of dark inclusions.

▶ A drill hole, from the surface of a faceted diamond, to reach an inclusion. FOV 1.66 mm.

A tiny hole 0.2 mm (0.008 in) wide is burned to reach the inclusion, starting from the table or crown facets to minimise the appearance of the hole when viewed from above. Very strong acid is then used to dissolve, or bleach, the inclusion making it less visible. Alternatively, a KM laser may be used to create small, repeated fractures in a zig-zag fashion to reach the inclusion, creating a more natural looking fracture than a drill hole.

Synthetic diamond was developed in the 1950s, becoming commercially available from the early 1990s. The hardness, inert nature and high thermal conductivity of diamond means synthetics have an incredible range of industrial uses, from abrasives, cutting tools, electronics, medical applications and lasers, to nanotechnology and quantum computing. As a faceted gemstone it is sold alongside natural diamond by many jewellers, showing the same brilliance and fire, but must be disclosed. The guidelines and nomenclature for synthetics are evolving constantly, and with their rise in popularity, it is important to stay up to date. The current nomenclature (*CIBJO The Diamond Book*, 2020) is synthetic, laboratory-grown or laboratory-created diamond, although it should be noted synthetic diamond is grown in factories, not gemmological laboratories.

Production of synthetic diamond has increased rapidly in the past decade, with the ability to create high-quality, colourless faceted gems over 15 ct, and mass produce melee, those tiny diamonds used as accent stones. Synthetic melee has been identified

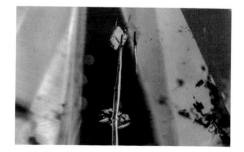

▲ A exceptional colourless, 10.02 ct emerald cut HPHT synthetic diamond.

The diamond grows from the dissolved carbon onto a seed crystal. In the past decade the production of HPHT diamond (including industrial diamonds) has exceeded 300 tonnes (330 tons) per year.

Different types of diamond can be grown by HPHT. Pure carbon creates colourless Type IIa stones, the addition of nitrogen creates yellow to orangish Type Ib, and the addition of boron creates blue Type IIb. Green diamond is grown through the addition of nickel, by combining nitrogen (yellow) and boron (blue), or post-growth irradiation treatment. The early diamonds were yellow to orangish, but since the mid-2000s colourless and blue have become more prevalent, with HPHT treatment used on the yellowish HPHT diamonds to decolourise them. Most melee is HPHT.

mixed with natural melee for several years, and the small size and volume make them challenging to distinguish. The detection of synthetics is understandably a dominant area of gem research. Some synthetic diamonds have features that can be detected by gemmologists, such as different inclusions, different responses to UV light, magnetism, and internal strain (or lack of) seen when using the polariscope. There are numerous testing devices on the market for colourless gems based on photoluminescence and UV light transparency. Gemmological laboratory examination is often needed for definitive identification.

HPHT diamond can be identified by several features. Most exhibit fluorescence, typically with a stronger reaction in shortwave UV light than longwave UV light (the opposite to natural diamond), and may show a diagnostic hourglass or cross-shaped fluorescence pattern. The longwave UV light response is yellow to orange, differing from the characteristic blue of many natural diamonds. HPHT diamond commonly phosphoresces following exposure to shortwave UV light. Some gemstones, especially early HPHT diamonds, are attracted to a magnet due to

There are two methods to synthesize diamond. The first recreates the high pressure, high temperature conditions of natural formation. Known as HPHT diamonds, the General Electric company announced the first successful growth by a reproducible method in 1955. The equipment used generates pressure by a split sphere multi-anvil press or a high-pressure belt apparatus, around a chamber in which the diamond is grown. The cell within the chamber contains industrial diamond powder or graphite (the source of carbon) dissolved in a molten metal alloy (flux). This allows diamond to grow at lower temperatures and pressures than required to transform graphite directly to diamond.

▼ Blue and yellow gem quality HPHT synthetic diamond crystals grown by General Electric in the 1980s.

needle-like metallic inclusions from the iron/nickel/cobalt flux in which they grew. Testing with a very strong rare earth element (REE) magnet can detect this. The identification of metallic inclusions, and lack of natural mineral inclusions, can distinguish synthetics. HPHT diamonds have clear growth sectors and are not internally strained thus will not show any anomalous extinction effects under cross polarization on the polariscope. This differs from natural diamond, which has complex growth and anomalous extinction from internal strain caused by its turbulent formation.

The second method used to synthesize diamond is chemical vapour deposition (CVD), first achieved in the 1950s, but growth of high-quality, single crystal layers for industry and gemstones did not begin until the late 1990s. It uses a chamber commonly made of glass, into which methane (CH_4) and hydrogen (H_2) are pumped. The methane is the source of the carbon, which deposits layer upon layer onto a growing surface (a synthetic or natural seed crystal) forming a tabular single crystal. The chamber has a moderate temperature but very low pressure (atmospheric pressure or less). The diamond is therefore unstable and oxygen is added during the growth process to etch away any graphite that forms. The CVD process initially produced thin layers of diamond, and wasn't seen on the market in gemstone sizes until around 2009. The size and quality have significantly improved over the past decade. In recent times, diamond simulants, such as synthetic moissanite and cubic zirconia, with a CVD diamond coating have appeared on the market. This gives the gemstone some characteristics of diamond, testing positive as diamond with a diamond tester, which measures the thermal conductivity of the surface of the stone. The strong double refraction of synthetic moissanite may be seen inside the stone under magnification.

CVD diamond has layered growth. When viewed with cross polarization it exhibits an anomalous columnar pattern from internal strain. It fluoresces but, unlike HPHT diamond, it has a short-lived or lack of phosphorescence. It may have inclusions of graphite particles that can look natural. It may also have silica impurities from the glass chamber, but does not contain any nitrogen impurities so is classed Type IIa. This distinguishes it from the majority (98%) of natural diamond which is Type I, but care must be taken to ensure it is not confused with natural Type IIa. In contrast to HPHT diamond, most CVD diamond is treated post growth. It tends to have a brownish colour, and most near-colourless CVD diamond has undergone HPHT treatment to remove brown hues. Anecdotal evidence suggests that CVD diamond may darken under ultraviolet light, or lighten under heat such as a jeweller's torch, however this effect is temporary.

Pink diamond cannot be synthetically grown, as the exact atomic level structural defect that produces the colour is still unknown. Synthetic pinks are created by irradiation of certain types of HPHT or CVD synthetic diamond followed by moderate heating.

Synthetic diamond may be graded by gem laboratories using the 4Cs, the same as natural diamond. The gemstones often have 'lab grown' or similar wording inscribed onto the girdle. Synthetic diamonds are an affordable alternative to natural diamonds and buyers like the appeal of a conflict-free diamond, with no mining impact on the environment. With rumours that unlimited carats could be produced, it will be interesting to follow this market in the coming years as the quantities and qualities continue to improve.

Colourless minerals such as zircon, topaz and beryl are used as diamond imitations, often backed by silver foil to increase the brilliance. Human-made imitations include glass, cubic zirconia, synthetic moissanite, strontium titanate, yttrium aluminium garnet (YAG) and gadolinium gallium garnet (GGG), some of which show similar fire and brilliance to diamond.

Cubic zirconia, often referred to as CZ, is a synthetic material developed in 1976 as a cheap simulant of diamond. CZ does occur naturally, but only as rare microscopic grains. It is zirconium oxide (ZrO_2) with a cubic crystal structure, so is singly refractive, the same as diamond. Its dispersion of 0.058–0.066 is high (higher than diamond) giving greater fire. Its refractive index is higher than most other gem minerals, although lower than diamond, enabling a bright vitreous lustre. As a sparkling fiery stone it makes a wonderful affordable alternative to diamond and is very common in costume jewellery. Cubic zirconia is normally transparent with few flaws and can be produced in any colour by the addition of metal oxides. Inclusions are commonly feather-like.

With a high hardness at 8.5, and no cleavage, cubic zirconia is a durable stone, resistant to scratching and fracture. However, it often appears more damaged after wear than other gems of lower hardness, perhaps as it is more brittle than diamond, and likely worn with less care due to its low value. Cubic zirconia has poor thermal conductivity, which means it will feel warm to the touch, in comparison to diamond's high thermal conductivity and cool feel. This property can be used to distinguish between the two, utilized in diamond testers, which measure the conduction of heat. Care must be taken as this also means it cannot withstand the same heat as diamonds. Cubic zirconia is not the same as the zirconium silicate mineral zircon ($ZrSiO_4$), also used as a diamond simulant for its fire.

Synthetic moissanite is a more recent material used to simulate diamond, commercially produced from 1998. It is a silicon carbide (SiC) and occurs naturally as a rare mineral in very small grains. All moissanite gems on the market are synthetic. Moissanite shares some of diamond's properties. It has a high hardness of 9.25–9.5, and is thermally conductive. Other properties exceed those of diamond, and its higher refractive index and very high dispersion of 0.104 means it has noticeably greater brilliance with more than twice the fire. It has a hexagonal crystal structure and exhibits strong double refraction, easily seen by a hand lens with a doubling effect of the back facet edges seen through the stone. This distinguishes it from singly refractive diamond. Moissanite generally has a slightly greenish-yellow colour, however in recent times has been treated to become colourless, and is graded colourless, near-colourless and faintly coloured. It may also be fancy coloured. Inclusions are not common, and may be thin parallel tubular features at right angles to the table of the faceted gemstones. Surface oxidation occurs over time, seen as an iridescent film, and this may be removed with certain polishes. Care should be taken when using diamond testers, as moissanite's thermal conductivity means it may test positive.

◀ 11.79 ct brilliant cut cubic zirconia showing high dispersion.

▼ A comparison of diamond and its imitations (L to R): quartz, sapphire, cubic zirconia, diamond and moissanite.

Diaspore

Composition AlO(OH) aluminium oxide
hydroxide **Crystal system** orthorhombic
Hardness 6.5–7 **Cleavage** perfect **Fracture**
conchoidal **Lustre** vitreous **SG** 3.20–3.50
RI 1.682–1.752 **Birefringence** 0.048 **Optical
nature** biaxial + **Dispersion** 0.0164

The beauty of diaspore resides in its colour, and
ability to change colour. Rare in gem quality,
it is sourced predominantly from one location
in Turkey and is sold under the trade names of
Zultanite, Csarite and Ottomanite, reflecting the
country's heritage. The name diaspore comes
from the Greek word 'diasperirein' which means
to scatter. This is due to its decrepitation or
tendency to break apart upon heating with a
blow pipe flame, a common way to analyze
minerals since the 1700s.

Diaspore is an aluminium oxide hydroxide, a
relatively common mineral and one of the main
mineral components of bauxite, the commercial
ore of aluminium. It normally occurs as lamellar
masses with a pearly lustre that are not of
interest to the gem cutter or collector, and large
gem quality crystals are rare. Diaspore may

also occur as a tiny mineral inclusion in ruby,
sapphire and spinel, minerals that are of similar
aluminium-rich compositions.

This mineral was first discovered in Russia
in 1801, and has since been found in many
other locations around the world, including
Argentina, Brazil, China, Myanmar, New Zealand,
Norway, South Africa, the USA and the UK.
Exceptional crystals have been found in the
Ilbir Mountains in southwest Turkey since the
late 1970s and this is the major source of all
gem-quality material. Here diaspore is mined
from metamorphic bauxite deposits, occurring
in veins of hydrothermal mineralisation within
fractures and cavities.

Diaspore is part of the orthorhombic crystal
system. Large gem crystals from Turkey are blocky
or elongate prismatic with square or rectangular
cross sections, and may be striated along the
length. Crystals are commonly found in V-shape
or dovetail twinning. They are transparent to
translucent and range from yellow through to
brown, green, grey, white, colourless to pinkish,
orange or red. Traces of iron and manganese
are the cause of the pastel colours. Diaspore is
strongly pleochroic displaying three colours: pale
yellowish green, brownish pink to red, and bluish.

◀ 6 cm (2.25 in) striated
diaspore crystal with a
3.4 ct gemstone, from
Muğla Province, Turkey.

◀ ▼ Pale brown 20.74 ct step cut diaspore, from Muğla Province, Turkey, appearing greenish in daylight (left) and pinkish under incandescent light (right).

Gemstones are cut to show flashes of the green and red pleochroic colours together as the stone is viewed table facet up.

Adding further beauty to this pleochroic effect, greenish brown gem-quality material exhibits a colour change from light olive-green in daylight, to golden under incandescent light, to light pinkish brown under candle light. These colour change specimens are the most desirable as faceted gemstones. The colour change effect is caused by impurities of chromium and iron.

While most common diaspore is very included, the gem-quality material from Turkey is often eye clean, with few inclusions visible under the hand lens. The bright sparkly nature of these gemstones with vitreous lustre, plus the pleochroism and colour change effect, make for quite spectacular gems. Very rarely stones cut en cabochon will show a chatoyant cat's eye optical effect due to many parallel needle-like inclusions.

Diaspore has a hardness of 6.5–7, so is reasonably resistant to scratches. However it is brittle with a perfect cleavage in one direction, and distinct in another, which reduces its durability, and care should be taken while wearing it. Diaspore can be cleaned using warm soapy water. Never use heat, steam or ultrasonic cleaners as this may damage the gemstone.

The cleavage makes diaspore a challenging stone to cut. The cutter must factor in the direction of cleavage and shape of the rough, while trying to capture the pleochroism and maximize the yield. Cut gemstones are often elongate to make the most of the prismatic crystals. Yield tends to be low, and large eye-clean colour change gemstones are very rare, demanding a premium price.

Varieties of diaspore include chrome diaspore, which is coloured light to dark purple by chromium, and mangan-diaspore, a rose-coloured manganese-bearing variety found in South Africa. An attractive purplish pink transparent diaspore was reported from Afghanistan in 2020, coloured by chromium and vanadium.

Colour change glass has been used as an imitation, containing rare earth elements or similar impurities to natural diaspore of manganese, chromium, iron and vanadium. A different colour change glass used to imitate alexandrite may also cause confusion. Both can be distinguished from natural diaspore using standard gemmological equipment.

Diopside

Composition CaMgSi$_2$O$_6$ calcium magnesium silicate **Crystal system** monoclinic **Hardness** 5–6.5 **Cleavage** good **Fracture** uneven, conchoidal **Lustre** vitreous **SG** 3.22–3.40 **RI** 1.664–1.730 **Birefringence** 0.024–0.033 **Optical nature** biaxial + **Dispersion** 0.017–0.020

Diopside is a common rock-forming mineral, typically occurring in a range of colours from colourless to green, brown or grey. As a gem it is best known as the attractive bright green variety chrome diopside. With a colour rivalling emerald, tsavorite garnet and chrome tourmaline, and a more affordable price, it is limited in use only by its low hardness. Other varieties include rare violet-blue violane, and black four-rayed star diopside. The name diopside comes from the Greek word 'diopsis' meaning double appearance or twice face, most likely referring to the shape of the crystal.

As a member of the pyroxene group of silicate minerals, and part of the clinopyroxene subgroup, diopside has monoclinic crystal symmetry. Crystals are prismatic with square cross sections and may be elongate or blocky, often creating attractive mineral specimens. It may also form granular masses. Diopside has two distinct cleavages at approximately 90 degrees, common to the pyroxene family. Diopside is a calcium magnesium silicate and forms a solid solution series with other clinopyroxene minerals, including hedenbergite (CaFeSi$_2$O$_6$), where iron substitutes for the magnesium. Almost all diopside contains some iron, becoming darker and more opaque with increasing iron content as it moves towards hedenbergite composition. Iron is the cause of green, yellow and brown colours.

Diopside is found in many different mafic and ultramafic igneous rocks, and metamorphic rocks such as marbles and skarns. Facetable material comes from numerous places including Italy, Austria, Madagascar, Myanmar, the USA, Canada, China, Brazil and Sri Lanka.

Chrome diopside predominantly forms deep in Earth's mantle in ultramafic peridotites, brought to the surface within fragments of mantle rock in kimberlite pipes, together with diamond and garnet. Chrome diopside is thus one of the minerals used as an indicator for diamond. The main source of gem-quality chrome diopside is the Inagli deposit in Siberia, Russia. This find in 1988 has made it commercially available, however the harsh weather and terrain means the deposit can only be mined during summer months. High-quality Siberian chrome diopside may be marketed as russalite®, imperial diopside and even the misleading term Siberian emerald. Other localities for chrome diopside include Myanmar, the Merelani Hills in Tanzania (found in association with tanzanite), diamond mines in South Africa, Pakistan, Afghanistan, Kenya, Australia, and fine crystals are known from Outokumpu in Finland.

◀ Prismatic diopside crystal from Sar-e Sang, Badakhshan, Afghanistan.

▲ Greyish 6.32 ct step cut diopside from Mogok, Myanmar.

▲ 3.12 ct oval mixed cut of rich green chrome diopside from Russia.

order to lighten the colour and improve the light return, giving more brilliance and sparkle. Chrome diopside is pleochroic exhibiting light green/yellow green/dark green. A rarer bright green variety coloured by vanadium is known as lavrovite, found in Italy and Russia.

The bright purple to pale bluish variety violane is coloured by manganese. Violane forms as granular material, very rarely as small crystals, in metamorphic rocks. It is mined in the Aosta Valley in Italy, Canada, Russia, and the USA and is quite rare. Violane is opaque to translucent with a coarse texture. It is used for carving ornamental objects, inlays or polished into cabochons. A mauve diopside coloured by titanium also exists but is different to violane.

Star diopside is typically opaque dark green to black. It is cut en cabochon to show the asterism caused by two sets of parallel inclusions, which produce a white four-rayed star. The 'arms' tend to be slightly wavy and are at oblique angles (around 105 and 75 degrees). The inclusions causing the asterism are commonly magnetite and the gemstones may be slightly magnetic – a convenient test of the identification! Star diopside will also have a slightly higher specific gravity due to these inclusions. Star diopside is primarily sourced from southern India. It is also found in Myanmar, which additionally produces chatoyant material exhibiting a sharp cat's eye effect caused by one set of inclusions.

Chrome diopside occurs in saturated vivid greens caused by traces of chromium impurities. It is transparent and often eye clean. Inclusions may be re-healed fractures and fluid inclusions. It is faceted to make best use of the long crystal form and to emphasize colour, with elongate step cuts, ovals and cushion cuts common. Gemstones tend to be small, most under 3 ct and rare over 10 ct, for a number of reasons. Diopside is challenging to cut due to the two cleavage planes; larger stones tend to be more included than smaller stones, and primarily because the saturated colour means larger stones become too dark, appearing almost black. Faceted gemstones are cut with shallow angles in

Diopside is relatively soft, so is used mainly in earrings and pendants, settings which offer more protection. The smaller size and bright colour of chrome diopside make it ideal as an accent stone in jewellery, adding a flash of green. Care should be taken when wearing, and it should be stored separately to other harder gemstones. Clean with warm soapy water and a soft cloth. Diopside is rarely treated. Synthetic diopside is known and has uses in the ceramic industry but is not normally seen on the gem market.

Feldspar Group

The feldspars are a group of closely related aluminium silicate minerals. Of these, only a few are desired as gems, including blue-green amazonite, and the varieties moonstone, labradorite and sunstone which are prized for their optical phenomena. Occasionally feldspars form transparent crystals which are faceted into attractive gemstones.

Feldspar

Composition aluminium silicate with K, Na, Ca **Crystal system** varies by species **Hardness** 6–6.5 **Cleavage** perfect **Fracture** uneven, conchoidal, splintery **Lustre** vitreous **SG** 2.54–2.72 **RI** 1.518–1.573 **Birefringence** 0.005–0.012 **Optical nature** biaxial + or - **Dispersion** 0.012

The feldspars are Earth's most common mineral group, collectively making up more than half of its crust. The major rock forming feldspars are the ones of most interest to the gem enthusiast. They are classified into two groups:

1. Alkali feldspars – a calcium-poor series of predominantly potassium-rich minerals, from the K-feldspars microcline, orthoclase and sanidine (polymorphs with the same composition $K(AlSi_3O_8)$ but forming under different temperature conditions), to sodium-rich albite $Na(AlSi_3O_8)$.

2. Plagioclase feldspars – a potassium-poor series from albite $Na(AlSi_3O_8)$ to calcium-rich anorthite $Ca(Al_2Si_2O_8)$.

These two groups form solid solution series between the potassium- to sodium-rich, and sodium- to calcium-rich end member compositions. Plagioclase feldspars are assigned names for the intermediate compositions between albite and anorthite. For example

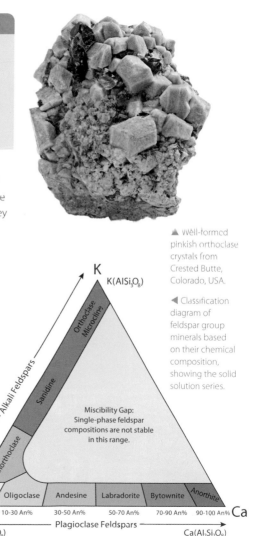

▲ Well-formed pinkish orthoclase crystals from Crested Butte, Colorado, USA.

◄ Classification diagram of feldspar group minerals based on their chemical composition, showing the solid solution series.

the popular labradorite is a plagioclase feldspar with a composition that falls within the ratios albite:anorthite of 50:50 to 30:70.

Feldspars occur in a range of deposits, from intrusive and extrusive igneous rocks, to metamorphic and sedimentary including secondary placer deposits. The gem feldspar species crystallize in monoclinic or triclinic symmetry. Crystals are commonly bladed, tabular or prismatic, and are typically twinned as both simple twins, and polysynthetic, or repeated twins, appearing as lamellae in the crystal.

As gem material, feldspar is sought for its wide-ranging optical effects and attractive colours. Moonstone and sunstone, named for their appearance, occur in more than one feldspar

species, so are discussed under gem variety. Less commonly, feldspar is seen as transparent faceted gemstones. These are pale with little fire due to the low dispersion, but their vitreous lustre makes them bright. Their fragile nature means they are not commonly seen on the market.

Feldspar is moderately hard at 6–6.5 with good to perfect cleavage in two directions at around 90 degrees, giving it poor toughness. It may scratch or chip easily so care should be taken to avoid knocks, and heat. Earrings and pendant settings provide protection, however feldspar can be used in rings that are not worn every day. Feldspar may be cleaned with warm soapy water, but avoid ultrasonic and steam cleaners.

Moonstone

Composition $K(AlSi_3O_8)$ orthoclase and sanidine, $(Na,Ca)[Al(Al,Si)Si_2O_8]$ oligoclase **Crystal system** monoclinic (orthoclase, sanidine), triclinic (oligoclase) **Hardness** 6–6.5 **Cleavage** perfect **Fracture** uneven, conchoidal **Lustre** vitreous **SG** 2.54–2.66 **RI** 1.518–1.547 **Birefringence** 0.005–0.010 **Optical nature** biaxial - (orthoclase, sanidine), biaxial +/- oligoclase **Dispersion** 0.012
(Rainbow moonstone – see labradorite, p110)

Moonstone is the most popular gem feldspar, named for its resemblance to moonlight. It shows a milky white, silver or bluish sheen or pale iridescent colours, which appear to move across the stone as it is turned. This optical phenomenon is named adularescence, also known as schiller.

The adularescence is caused by sub-microscopic intergrowths of two different types of feldspar. The crystal was originally a blend between these two feldspar

▼ A 33.7 ct high domed cabochon of orthoclase moonstone from Brazil, with bluish adularescent sheen.

compositions, but as it cooled they separated out into alternating layers of varying thickness, in a process known as exsolution. These intergrowths are similar in scale to the wavelengths of visible light. The optical effect occurs as light waves are reflected from the internal layered structure. The waves interfere to produce iridescence or, where the

▲ 'Centipede' inclusions of tiny intersecting stress cracks, in orthoclase moonstone. FOV 2.01 mm.

intergrowths are smaller than the wavelengths of light, are scattered. Blue wavelengths of light are scattered more than other colours, so the sheen appears bluish.

Only certain compositions of feldspar exsolve into the microscopic layering, therefore only certain gem feldspars exhibit adularescence. The physical and optical properties of moonstone consequently vary by composition. Most moonstone is a variety of orthoclase containing a small portion of albite. This separates into potassium-rich and sodium-rich layers. An adularescent sanidine is known from the USA. Oligoclase shows adularescence via exsolved lamellar intergrowths of oligoclase and albite. This often exhibits a subtle iridescence and is also known as peristerite. Rainbow moonstone is a labradorite feldspar with a darker transparent body colour and multi-coloured adularescence, found in India and Madagascar.

Moonstone is transparent to opaque and has a range of body colours including colourless, white, grey, yellow, orange, red, brown or green. The yellow to brown colours are caused by inclusions of iron oxides. The most valued moonstone is transparent and colourless with a strong blue sheen. While transparent moonstone is highly desirable, translucent material can emphasize the adularescence. Moonstone has diagnostic 'centipede' inclusions of tiny stress cracks. These are fissures intersected by many short cracks at 90 degrees, aligned parallel to the cleavage directions, in a pattern resembling a centipede. Moonstone can occasionally have a chatoyant cat's eye effect and very rarely a four-rayed star.

The most important source of moonstone is Sri Lanka, where it is mined from weathered pegmatites and gem gravel placer deposits. Other sources include India, Madagascar, Myanmar (historically important), Brazil, Tanzania, Australia, the USA, Canada, Russia, Norway, Austria and Switzerland. India produces moonstone of orange to brownish body colour including cat's eye and star varieties. Myanmar is also known for chatoyant cat's eyes caused by inclusions of parallel needles.

The adularescent effect is best seen in moderately domed cabochons, and moonstone is typically fashioned this way. Oval shapes are common. Cabochons also reveal any chatoyant cat's eye effect or asterism. Moonstone is additionally faceted or fashioned into beads. It may be carved and cameos are popular, particularly of moon faces. Moonstone occurs in a range of sizes, although high-quality gemstones generally weigh less than 15 ct. Moonstone was fashionable in the early 1900s, favoured in art nouveau jewellery, and it remains popular as a widely available and affordable gemstone.

Moonstone is not normally treated, but on occasion is impregnated, or surface-coated for durability. Coatings on the back of the stone have been used to enhance the adularescence, or impart a body colour or cat's eye effect. Moonstone may be imitated by synthetic spinel with an adularescence, milky chalcedony or quartz, and opalescent glass.

Labradorite

Composition (Ca,Na)[Al(Al,Si)Si$_2$O$_8$] plagioclase feldspar **Crystal system** triclinic **Hardness** 6–6.5 **Cleavage** perfect **Fracture** uneven, splintery **Lustre** vitreous **SG** 2.68–2.72 **RI** 1.554–1.573 **Birefringence** 0.007–0.012 **Optical nature** biaxial + **Dispersion** 0.012

Labradorite is a popular gemstone known for its iridescent colours. This optical phenomenon is named labradorescence, also known as schiller. Labradorite is a plagioclase feldspar with a compositional range albite:anorthite of 50:50 to 30:70. As its composition is generally closer to anorthite, it is sometimes considered a sodium-rich variety of anorthite.

This gem is named after Labrador, Canada, where it was discovered. It is known from many locations, found in igneous and metamorphic rocks. Crystals are tabular and almost always twinned, however most labradorite occurs as massive material or intergrown aggregates. Significant gem sources are Canada, Madagascar, Finland, Russia, Australia, the USA and Mexico. Dark material with an intense range of iridescent colours (sourced mostly from Finland) is sold under the trade name spectrolite. Larvikite is an attractive bluish-grey building stone, predominantly sourced from Norway, composed of interlocking labradorite crystals with a schiller effect.

Labradorite is translucent to opaque with grey to yellowish body colour. It typically contains needle-like ilmenite, and sometimes platy magnetite inclusions, which cause a darker body colour. The magnificent iridescence of bright metallic spectral colours is revealed when the stone is viewed from certain angles. Blue to green iridescence is typical, but it can exhibit the full spectrum with yellow, bronze, red and purple iridescence the rarest. A transparent labradorite with a range of iridescent colours is marketed as rainbow moonstone.

▼ Labradorite specimen with a polished face showing exceptional iridescence, from Labrador, Canada.

▼ A cameo carved in labradorite with blue iridescence.

The cause of the optical phenomenon is complex, involving diffraction and thin film interference. Similar to moonstone, the albite-anorthite compositional blend separates during cooling into microscopic parallel intergrowths of sodium- and calcium-rich layers. The lamellae alternate thin (albite) and thick (anorthite) with a thickness less than the wavelengths of visible light. Light is reflected and refracted from these thin films, and also diffracted, interfering to cause the spectral colours. As the anorthite content increases, so does the lamellae thickness, causing the iridescence to contain more orange and red.

Labradorite is fashioned into cabochons in a range of shapes with low domes, to best reveal the labradorescence. It is also used for beads, and more translucent material may be faceted. The rough can occur in large sizes and is carved, or polished into freeform decorative shapes, however the perfect cleavage makes it a challenge to fashion. Finished gems may also be large,

commonly several centimetres wide, creating very attractive pendants, earrings and even rings.

On rare occasions labradorite occurs as transparent gem-quality material without labradorescence. This is found in the USA, France, Australia and Mexico, and faceted into gemstones of pale yellowish colours.

Sunstone

Composition (Na,Ca)[Al(Al,Si)Si$_2$O$_8$] oligoclase, K(AlSi$_3$O$_8$) orthoclase **Crystal system** triclinic (oligoclase), monoclinic (orthoclase) **Hardness** 6–6.5 **Cleavage** perfect **Fracture** uneven **Lustre** vitreous **SG** 2.54–2.67 **RI** 1.518–1.552 **Birefringence** 0.005–0.010 **Optical nature** biaxial +/- (oligoclase), biaxial - (orthoclase) **Dispersion** 0.012
(Oregon sunstone – see labradorite, p110)

▶ Oligoclase sunstone from Malawi. The aventurescence is caused by hematite inclusions.

Sunstone is a warm reddish feldspar with a shimmery optical phenomenon known as aventurescence. This is also sometimes called schiller. Aventurescence is caused by inclusions of tiny metallic platelets, all in the same orientation, which reflect light. Sunstone was originally named as a variety of oligoclase feldspar containing reddish hematite inclusions, but is now applied to any aventurescent feldspar.

This gemstone ranges from colourless through yellow, orange, brown, to occasionally green. It is transparent to opaque. The inclusions may be iron oxides of hematite or goethite, copper or another mineral. The inclusions influence the body colour with hematite imparting reddish colours, and copper giving orangey warmth and the rarer cool greens. The sizes of the inclusions define the aventurescence. Large ones give a glittery effect, smaller ones create a sheen, and if too small to be seen by eye, they may only

▼ Some of the beautiful colours of Oregon sunstone.

impart a colour without aventurescence. Rarely sunstone exhibits a chatoyant cat's eye effect or asterism.

Sunstone occurs as crystals in pegmatites, metamorphic rocks and as phenocrysts (distinctly larger crystals) in volcanic rocks. Sources include India, Norway, Russia, Madagascar, Australia, Tanzania, Canada, Mexico and the USA (Oregon).

▶ Goldstone glass containing triangular copper inclusions, creating aventurescence to imitate sunstone.

Oregon sunstone is a transparent plagioclase feldspar with inclusions of copper. It was named the official state gemstone of Oregon in 1987, and is gaining in popularity. Its composition falls mostly within the labradorite feldspar range, and colour varies from colourless, pale yellow, orange and red, to brown and green. Deep reds are known as spinel red, and these together with green and multicolour gemstones are the most desirable. Oregon sunstone does not show aventurescence if the copper inclusions are too small.

This feldspar occurs as phenocrysts trapped within old lava flows, and the often fractured crystals are extracted from the weathered basalt rocks. The fragments are commonly colour zoned, and cutting of the rough is selective to create gems of one or multiple colours. Oregon sunstone is not treated, adding to its appeal.

Rainbow lattice sunstone is the trade name for an orthoclase feldspar found in the Harts Range, Australia. It contains hexagonal platelet inclusions of hematite and ilmenite, forming an iridescent lattice along intersecting cleavage planes. It may also exhibit a moonstone adularescence.

Sunstone is faceted, carved or polished into cabochons in a range of sizes, often with low domes to better show the aventurescence. Fancy and fantasy cuts with unusual shapes are popular for transparent Oregon sunstone, making use of the larger rough, the shimmer and especially to emphasize multiple colours. Oregon sunstone is also cut as calibrated gemstones. Sunstone is simulated by aventurine glass or goldstone, an artificial glass with copper inclusions. This is said to have been accidentally created in the 1600s in Italy, and the names aventurine and aventurescence are derived from the Italian 'a ventura' meaning by chance. It can be distinguished by the opaque uniform colour, and under magnification by the triangular to hexagonal shape of the copper inclusions.

In the early 2000s, a copper-bearing andesine feldspar of attractive uniform red, and less commonly green colour, appeared on the gem market, reported to be from China or the Democratic Republic of the Congo. Great controversy has surrounded this gemstone and the source. It has since been shown to be a treated pale yellow andesine diffused with copper under high temperature to create the uniform red colour.

Amazonite

Composition $K(AlSi_3O_8)$ potassium aluminium silicate **Crystal system** triclinic **Hardness** 6–6.5 **Cleavage** perfect **Fracture** uneven, conchoidal **Lustre** vitreous **SG** 2.54–2.63 **RI** 1.522–1.530 **Birefringence** 0.008 **Optical nature** biaxial - **Dispersion** 0.012

Amazonite, or amazonstone, is the greenish blue to green variety of K-feldspar (usually microcline but may be orthoclase). It is named after the Amazon River along which it was reputedly first found, however this source has never been verified.

The cause of the attractive greenish colour is thought to be a combination of lead impurities,

▼ Amazonite crystals on smoky quartz, from Smoky Hawk claim, Teller Co., Colorado, USA.

Colorado, it occurs in spectacular combinations with dark smoky quartz, which are sought by mineral collectors.

Opaque to translucent, amazonite has a vitreous lustre. It is typically fashioned into cabochons to highlight the colour, which also offer a protective shape. It may be used in beads, carved and, on rare occasions, translucent material is faceted. Inclusions of incipient cleavage can cause an appealing shimmery effect, but are an indication of low durability. Avoid knocks due to the perfect cleavage, brittle nature, and possible internal fractures. It is recommended to wear amazonite in more protected settings.

Although not normally treated, amazonite may be dyed to improve or even out the colour. It may also be coated or impregnated with wax or polymers to improve the lustre, hardness and durability. Glass and similarly coloured chalcedony, such as chrysoprase, are used to imitate amazonite. It may be confused with jade and turquoise, however jade lacks the white streaks and has a higher lustre, and turquoise lacks cleavage.

water within the crystal structure, radiation and possibly iron. Interestingly, exposure to sunlight can sometimes enhance the colour, which is the opposite of many other gemstones that will fade. The colour may be uneven, often with a white grid-like or parallel streaky pattern caused by the presence of albite feldspar. Uniform deep colour is the most desirable.

Amazonite is found in granite and granitic pegmatites. It has triclinic crystal symmetry, occurring as well-formed prismatic crystals, although often containing fractures. Sources of gem-quality material include the USA, India, Peru, Brazil, Madagascar, Kenya, Namibia, Myanmar, Russia, Afghanistan, Canada and Australia. In some locations, such as the Rocky Mountains,

▼ A polished piece of amazonite of rich greenish blue with streaky white texture.

Other non-phenomenal feldspar

Orthoclase is a common feldspar normally opaque white to pinkish, and rarely occurring in water-clear colourless to pale yellow gem quality. The name orthoclase is derived from the Greek 'orthos' meaning straight or right and 'klásis' meaning break, in reference to the perfect cleavage at right angles. Crystals are prismatic and often twinned and this was historically used to differentiate orthoclase from microcline. Orthoclase has a hardness of 6 and is the mineral used to define 6 on the Mohs scale of hardness.

▲ A large 29.95 ct fancy cut 'orthoclase' gemstone from Madagascar.

Orthoclase is found in granite, syenite and felsic volcanic rocks. Colourless water-clear crystals are known as the variety adularia, named after the Adular Mountains in the Swiss Alps where it was discovered. Adularia often shows adularescence, which is named after it. It forms in lower temperature geological settings and is sourced in Switzerland, Austria and the gem gravels of Sri Lanka and Myanmar. Pale yellow orthoclase is known from Kenya and Myanmar.

Magnificent transparent yellow crystals of K-feldspar have been found over the past century near Itrongay, Madagascar. Originally recorded as orthoclase, research has since indicated this is actually sanidine. Well-formed, undamaged crystals are sought by mineral collectors, and gem rough is faceted into attractive gemstones. The yellow colour is caused by impurities of iron. The crystals formed where pegmatites intersect marble. Sanidine was first found in Germany, and is also known from Sri Lanka, Myanmar, the USA and Mexico as colourless, pale brown or smoky gemstones. It may also show the adularescent effect of moonstone.

Oligoclase has triclinic symmetry forming short prismatic to tabular crystals, which are often twinned. It is found in granitic pegmatites and metamorphic rocks, occasionally forming colourless to bluish-green transparent gems. Sources include Brazil, Kenya, Tanzania and Madagascar.

Albite is infrequently used to create colourless or pale bluish faceted gemstones, and the name is derived from the Latin 'albus' meaning white, for its typical colour.

◀ Smoky grey 1.27 ct step cut sanidine from Eifel, Laacher See, Germany.

Fluorite

Composition CaF$_2$ calcium fluoride **Crystal system** isometric **Hardness** 4 **Cleavage** perfect **Fracture** splintery **Lustre** vitreous **SG** 3.00–3.25 **RI** 1.428–1.448 **Birefringence** none **Optical nature** Isotropic **Dispersion** 0.007

▶ Magnificent blue cubic crystals of fluorite.

Fluorite is one of the most colourful minerals, occurring naturally in a wide range of hues, often with multiple colour zones, creating very attractive gemstones. Due to its low hardness and perfect cleavage, it is not commonly used in jewellery and is faceted as a collector's gemstone. The rare variety Blue John, mined only in the Castleton area of Derbyshire, England, is used as a decorative stone for its purple, yellow, and white banding.

▼ Step cut 35.78 ct fluorite from Rotherhope Fell Mine, Cumbria, England.

Originally known as fluorspar, fluorite gained its name from the Latin verb 'fluere' meaning to flow, from its use as a flux in metal smelting. Known for many years prior, it was officially named in 1797, and gave its name in 1852 to the phenomenon of fluorescence and, in 1886, to its constituent element fluorine. It is the main ore of fluorine, which has many uses. Examples include the addition of the fluoride ion into drinking water and toothpaste to help prevent tooth decay, and as a fluorine polymer to make non-stick surfaces such as Teflon. Fluorite has a low refractive index and low dispersion, which means as a gemstone it can lack sparkle. These properties however make it of great use in optical lenses such as in cameras.

Fluorite occurs in hydrothermal veins, often associated with lead and zinc minerals. It is also found in granite and granitic pegmatites, and metamorphic rocks such as marble. It is a common mineral worldwide with large deposits in China, South Africa and Mexico. Other sources include Argentina, Austria, Canada, England, France, Germany, India, Morocco, Myanmar, Namibia, Russia, Spain, Switzerland and the USA. Highly desirable collector specimens include pinks from the Alpine regions of France, blues from Cave-in-Rock, Illinois, USA, water-clear crystals from Dal'negorsk in Russia and specimens from Peru, Berbes in Spain, Cornwall and the north of England.

Fluorite is a calcium fluoride and crystallises in the isometric system, typically forming as cubes or octahedrons. Complex variations of the two are not uncommon, and rarely it is fibrous or even globular. Crystals may show stepped surface growth or bevelled edges. Fluorite forms in groups of crystals, and twinning is common.

Gemstones range from transparent to opaque with a vitreous lustre. Fluorite has perfect cleavage in four directions, orientated on the octahedral planes of the crystal. This is seen externally as cleaved faces, which may look pearly or show iridescence, and internally as mirror-like flat planes of incipient

▲ Fluorite split along the four cleavage planes to create octahedrons.

cleavage. Other common inclusions are rehealed fractures, negative crystals, fluid inclusions, two- or three-phase inclusions, or solid crystals of other minerals. Fluorite ranges from colourless to almost every colour of the rainbow. Most common are purples, greens, yellows and blues. A single crystal may be one or more colours, exhibiting zoning or spotting aligned to the crystal form. The causes of colour are variable, hence the great variety, and include impurities (commonly rare earth elements), and imperfections in the crystal structure caused by natural radiation. Not all colours are stable, and some will fade over time if exposed to light. Being isometric, it does not show pleochroism or double refraction. Colour change fluorite is rare but is known from several locations. From India it ranges from brown in daylight to lilac in incandescent light. Brazilian fluorite containing rare earth element impurities will change from blue resembling topaz in daylight to purple resembling amethyst in incandescent light.

The majority of fluorite fluoresces under ultraviolet light, some with marked reactions, and it is the original mineral in which fluorescence was first observed. It can fluoresce in almost any colour under longwave ultraviolet, with blue, purple or green the most common, and may show a different reaction under shortwave UV light. The fluorescence is often caused by impurities of rare earth elements such as erbium, yttrium and europium. Some specimens exhibit daylight fluorescence, where the phenomenon may be seen outside or near windows, caused by the UV light contained in natural daylight. Emerald green specimens from the north of England are famed for their notable violet-blue daylight fluorescence. Fluorite may also show phosphorescence (continuing to emit light when the UV light source is removed), triboluminescence (emitting light after being struck or rubbed) and thermoluminescence (emitting light when heated).

Fluorite has a low hardness of 4, so may be easily scratched. It is brittle, which combined with the perfect octahedral cleavage makes it a challenging gem material to cut. Transparent rough may be faceted, and translucent to opaque

▲ Green fluorite from The Gulf, Emmaville, NSW, Australia with a 210 ct step cut gemstone (centre).

▲ Intense violet blue fluorescence under longwave UV light.

▲ Strong fluorescence under shortwave UV light reveals unexpected banding.

◀ A 45 ct fluorite gemstone with attractive colour banding from near Durango, Mexico.

▶ A magnificent vase carved from Blue John fluorite, 76 cm (30 in) in height, thought to be the largest Blue John vase in existence.

material is often made into cabochons or beads. The cleavage is also used to advantage to create attractive octahedral-shaped 'crystals' from massive transparent material. Gemstones are cut to make the most of the colour, and clever use of the colour zoning creates attractive gems.

Due to its softness and low durability, fluorite should always be treated with care, especially if pieces are used for adornment. Clean gently with warm soapy water, but never use heat, steam or ultrasonic methods. Store out of sunlight to protect the colour from fading.

Irradiation of colourless or pale green fluorite may produce beautiful purplish blue colours. Gentle heating of fluorite including the Blue John variety will lighten its colour, but care must be taken as higher temperatures of 200–300°C (392–572°F) will remove the colour completely. Synthetic fluorite is known in all colours, however as this fragile mineral is not commonly used as a gemstone, its use is mostly in optical lenses.

Blue John is a rare variety of fluorite from Castleton in Derbyshire, England. This beautiful stone is formed of deep purple to bluish purple, white and yellow bands. It has been mined since at least the start of the 1700s and used as a decorative stone for centuries. The fluorite occurs as banded veins with layers of interpenetrating cubes. When cut and polished in sections, the layering of the cubic crystal faces gives a zigzag effect to the band interfaces. The Blue John

forms from hydrothermal fluids within a fractured limestone, crystallizing in successive layers as veins and void fillings.

There are several theories behind the naming of Blue John. One is that it was named from the French 'bleu et jaune' meaning blue and yellow. The cause of the dark bluish to purple colours is due to imperfections in the crystal structure caused by low levels of natural radiation from surrounding rocks. Unlike other fluorites, Blue John does not fluoresce as it lacks rare earth element impurities. Blue John is normally included with fluid inclusions, fractures and incipient cleavage. Once mined, Blue John is impregnated with resin before it can be worked, to fill fractures and cleavages, bonding and improving clarity. It can then be carved into vases, bowls or ornaments, or used as inlaid stone, jewellery or other decorative items. Today artisans continue their work in Castleton, predominantly creating jewellery and small ornaments. Large historical items such as vases demand high prices. While a similar banded fluorite does occur elsewhere in the world, such as China, the colour and banding of Blue John is unique to Derbyshire.

Garnet Group

Garnets are a group of silicate minerals found all over the world, and are among the oldest known gemstones. Historically, garnet is associated with red, but this diverse group occurs in almost all hues including pink, purple, orange, yellow, green and colour change varieties. Today many colours are popular, with bright greens the most valued.

Garnet

Composition group of silicate minerals
Crystal system isometric **Hardness** 6.5–7.5
Cleavage none **Fracture** conchoidal **Lustre** vitreous, subadamantine **SG** 3.5–4.3
RI 1.714–1.940 **Birefringence** none **Optical nature** isotropic **Dispersion** 0.020–0.057

Red garnets have been prized for more than 5,000 years, found in the jewellery of Egyptian pharaohs and ancient Romans, through to royalty in Victorian times. They were seen as a symbol of life, and worn into battle as a talisman of victory and protection.

The name garnet is thought to be derived from the medieval Latin 'granatus' meaning pomegranate (as the deep red, rounded crystals resembled the seeds) and the 14th-century Middle English word 'gernet' meaning dark red. The term 'anthrax', meaning coal, was used by ancient Greeks, alluding to the gem's dark red glow. In Roman times the word 'carbunculus' was used for red gems including garnet, and the term carbuncle is still associated with dark red garnets today.

The garnet group comprises closely related mineral species with the same crystal structure, similar properties and a range of compositions. Each species is an end member and forms a solid solution series to other end member compositions, with elements of similar sizes substituting for each other in the crystal structure. The generic chemical formula for the garnets can be written $A_3B_2(SiO_4)_3$ where the A is filled by iron, magnesium, calcium or manganese and B is filled predominantly by aluminium, chromium or iron. Garnets are rarely found in nature in their pure end member compositions, but as endless blends between. This compositional variation gives the garnet group its diversity in properties such as colour, however it makes individual garnets difficult to classify as each crystal can be a mixture of two or more different species. As the composition shifts between species with heavier elements replacing lighter ones, the properties of specific gravity and refractive index increase.

The garnet group currently consists of 14 species, of which six are used as gemstones: almandine, pyrope, spessartine, andradite, grossular and uvarovite. These six are divided into two groups based on continuous solid solution series. The first group has aluminium in the B site, known as pyralspite for **pyr**ope, **al**mandine and **sp**essartine. These garnets can be any blend of the three end members. The second has calcium in the A site – ugrandite for **u**varovite, **gr**ossular, and **and**radite.

As chemical analysis is required to determine the exact composition (often requiring destruction of the sample), classification by gemmologists is defined using a combination of refractive index, specific gravity, absorption spectra, and more recently magnetism. This is normally portioned into end members (around 70% or more of one composition) or intermediate series. Additionally, and perhaps adding more confusion, many varietal and trade names exist on the market based on colour and source.

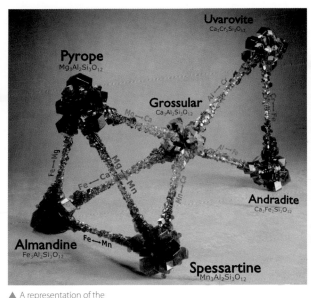

Uvarovite
$Ca_3Cr_2Si_3O_{12}$

Pyrope
$Mg_3Al_2Si_3O_{12}$

Grossular
$Ca_3Al_2Si_3O_{12}$

Andradite
$Ca_3Fe_2Si_3O_{12}$

Almandine
$Fe_3Al_2Si_3O_{12}$

Spessartine
$Mn_3Al_2Si_3O_{12}$

▲ A representation of the gem garnets by species, showing their chemical composition, interconnection and typical colour.

▼ A trapezohedral garnet crystal from Utah, USA.

▶ This spessartine is a complex combination of crystal forms, on white albite from Pakistan.

Garnets are found throughout the world, most commonly in metamorphic rocks, but also igneous and sedimentary. They are durable so may be weathered out from their hosts and deposited in secondary alluvial placer deposits. They are recorded from Ancient Carthage, Africa and India as far back as 2,500 years ago. Bohemia (now part of the Czech Republic) was the primary source of the spectacular deep red garnets favoured in Victorian jewellery. The Ural Mountains in Russia provided the green andradite variety demantoid prized by the Russian royal family and still desired today. The African continent is now the main source of garnets in a range of colours and rarer varieties, with many mined by small-scale artisanal miners. Other sources include Australia, India, Myanmar, Brazil, Sri Lanka, Iran, Afghanistan and Pakistan.

Crystallizing in the isometric crystal system, garnets predominantly form as dodecahedrons,

trapezohedrons, or a combination of the two. Although they are singly refractive, under the polariscope they may show anomalous extinction effects due to internal strain in the crystal structure. This is often caused by the high pressure and temperature conditions under which they formed.

Garnets range from transparent to opaque with a range of inclusions. Gem-quality material is faceted, and lesser material polished into cabochons. Some garnets, such as almandine, have such a deep red body colour that they are cut in a hollow cabochon with a concave back, to let more light pass through the stone. Rarely, a four- or six-rayed star effect occurs when cut en cabochon, caused by inclusions of rutile or elongated voids orientated in several directions.

Garnets have a bright vitreous lustre and medium to high dispersion, which creates very lively

▼ Almandine cabochons cut with concave backs to lighten the deep red colour, commonly referred to as carbuncles.

gemstones, especially in paler colours. Some garnets may exhibit a colour change effect similar to alexandrite, appearing bluish green in daylight and purplish red in incandescent light.

As durable gemstones with good hardness and a lack of cleavage, garnets are suitable for most jewellery. The variety demantoid is the softest garnet, and brittle, so care should be taken with this stone. Garnets are safe to clean with warm soapy water, and can be used in ultrasonic cleaners except when highly fractured or included. They are heat sensitive, so sudden changes in temperature should be avoided to prevent fracturing – avoid steam cleaners. Although they are resistant to chemical attack, it is best to avoid exposing them to anything acidic. Garnets are rarely treated, which adds to their appeal. They may have surface-reaching fractures filled to improve clarity and durability. Demantoid and occasionally hessonite may be heated to improve colour.

Two human-made garnets are known and are unique members of the wider garnet group. They have no known natural counterparts, so although they are often called synthetic garnets, they are correctly termed artificial. Both have been used as gemstone simulants, particularly to imitate

diamond. Yttrium aluminium garnet (YAG) was the first simulant, and became readily available in the 1970s. Colourless YAG was used as a diamond imitation, but it can be made in any colour and is still found today. Gadolinium gallium garnet (GGG) has a high dispersion (0.038–0.045) close to diamond. It can similarly be made in any colour. However, the advent of cheaper cubic zirconia means neither are commonly seen on the gem market.

Garnets are imitated by glass. They can themselves be used to imitate other gemstones in the form of a 'garnet-topped doublet' (GTD). This is a slice of natural garnet fused to glass, then faceted with the garnet as the table. The colour of the glass will determine the overall colour of the gemstone, regardless of garnet colour (which is usually red). The garnet top means the gemstone is more durable, with sharper facet edges and higher lustre than can be achieved from glass. Reflected light off the surface may reveal the join line, however jewellery settings may sometimes hide the join. Garnet-topped doublets have largely gone from use since the invention of synthetic gems, but may be found in antique jewellery.

▼ Green glass topped by a thin slice of red garnet, but appearing green face up. Arrows indicate the join.

Almandine

Composition $Fe_3Al_2(SiO_4)_3$ iron aluminium silicate **Crystal system** isometric **Hardness** 7–7.5 **Cleavage** none **Fracture** sub-conchoidal **Lustre** vitreous **SG** 3.88–4.30 **RI** 1.770–1.830 **Birefringence** none **Optical nature** isotropic **Dispersion** 0.027

Almandine is the most common garnet in the world. Dark red to purplish, it is often darker and browner than pyrope and may even appear black due to its depth of colour. Lighter colours include pinkish red or orangey red, and mauve. Almandine ranges from transparent to opaque, with characteristic inclusions of rutile needles and rounded crystals of other minerals such as apatite, or zircons with tension cracks or halos. When rutile needles are present in enough quantity, and the stone is cut en cabochon, it may exhibit asterism with four-, six- or 12-rayed stars. Gem-quality almandine is faceted, with more included material made into cabochons or beads. Very dark red transparent material is cut as a cabochon with a hollow concave back, to reduce the thickness and lighten the colour. Due to its iron content, it is attracted to strong neodymium magnets.

This garnet has a characteristic absorption spectrum from iron, which is the cause of the red colour. This, however, is not diagnostic of the end member species, as many garnets are a blend of species. It only indicates the garnet contains iron-rich almandine.

▼ Almandine crystals in matrix from Långban, Sweden.

Almandine occurs worldwide, primarily in metamorphic rocks such as mica schist and gneiss, as well as igneous rocks including pegmatite. Due to its resistance to weathering, gem-quality stones are found in alluvial placer deposits. Important sources include Brazil, India, Madagascar, Myanmar, Sri Lanka, Tanzania, the USA and Zambia.

Almandine forms a solid solution series with pyrope (iron substituting for magnesium) and spessartine (iron replaced by manganese).

◄ 16.44 ct faceted almandine garnet from Russia.

Pyrope

Composition Mg$_3$Al$_2$(SiO$_4$)$_3$ magnesium aluminium silicate **Crystal system** isometric **Hardness** 7–7.5 **Cleavage** none **Fracture** conchoidal **Lustre** vitreous **SG** 3.62–3.87 **RI** 1.714–1.766 **Birefringence** none **Optical nature** isotropic **Dispersion** 0.022

▲ Cushion mixed cut pyrope garnet of 11.17 ct.

The name pyrope comes from the Greek 'pyropos' meaning fiery alluding to its orangey red to red colours. It is famous from Bohemia (now part of the Czech Republic), the main source of garnets in Europe from the 1500s, and was popular in the 1700s to 1800s, especially in the Victorian era. It may still be marketed under the name of Bohemian garnet.

Pyrope is often eye clean with few inclusions, which may be needle-like or rounded crystals. It has a high refractive index and forms brilliant, lively faceted gemstones, although this may be masked by deeper colours.

Unlike other garnets, pyrope is more commonly of igneous rather than metamorphic origin. It forms deep in the Earth in ultramafic igneous rocks, such as peridotite in Earth's mantle. It is brought to the surface in kimberlite pipes, and is one of the minerals used to indicate the presence of diamond-bearing rocks. It is also found in high-pressure metamorphic rocks, and alluvial placer deposits.

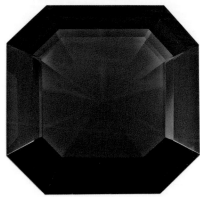

▲ 24.50 ct octagonal step cut pyrope from Mogok, Myanmar.

Pyrope forms a solid solution series with almandine, substituting magnesium for iron. Pure pyrope is actually colourless; red pyrope thus contains some almandine as iron causes the colour, and therefore it will also show an 'almandine' spectrum. Impurities of chromium create a richer red, as seen in fine Bohemian garnet and 'ant-hill' garnet from Arizona, USA. Pyrope is also attracted to strong neodymium magnets due to its magnesium content.

Some pyrope-almandine may show asterism when cut en cabochon from inclusions of rutile needles.

Little pyrope is produced in the Czech Republic today. Notable sources include South Africa, Tanzania, Mozambique, Kenya, India, Sri Lanka, the USA, Argentina, Australia, Brazil, Myanmar, Scotland, Switzerland, Russia and China. Interestingly it has also been found as inclusions within diamonds due to their associated formation.

Spessartine

Composition $Mn_3Al_2(SiO_4)_3$ manganese aluminium silicate **Crystal system** isometric **Hardness** 7–7.5 **Cleavage** none **Fracture** conchoidal **Lustre** vitreous **SG** 4.00–4.25 **RI** 1.789–1.820 **Birefringence** none **Optical nature** isotropic **Dispersion** 0.027

Spessartine (also called spessartite) is coloured by its manganese content, ranging from yellow, dark orange to red or brown with increasing iron content. Most desired is bright orange with fiery orange gems marketed under the name mandarin garnet. Not as common in gem quality, spessartine tends to have eye-visible inclusions such as wavy feathers, liquid inclusions or rutile needles.

▲ Vibrant orange cushion mixed cut spessartine weighing 6.22 ct.

It forms a solid solution series with almandine, which influences the colour, and most red or darker gemstones give the typical almandine spectrum. Less often, spessartine is blended with grossular, or pyrope which sometimes exhibits a colour change.

Spessartine is named after the Spessart district of Bavaria, Germany, where deposits were first found in the early 1800s. Spessartine occurs worldwide, primarily in granitic pegmatites and alluvial deposits, but is also found in metamorphic rocks and skarns. Sources include Sri Lanka, Madagascar, Australia, Brazil, Sweden, Myanmar, the USA and Tanzania. Mandarin garnet was first found in Namibia in 1991, later in Nigeria. Spectacular lustrous red crystals with stepped/etched growth from the Navegadora claim in Minas Gerais, Brazil, are sought by mineral collectors.

▶ Spessartine garnet with schorl, on matrix, from Little Three Mine, California, USA.

Andradite

Composition $Ca_3Fe_2(SiO_4)_3$ calcium iron silicate **Crystal system** isometric **Hardness** 6.5–7.5 **Cleavage** none **Fracture** conchoidal **Lustre** subadamantine, vitreous **SG** 3.70–4.10 **RI** 1.855–1.940 **Birefringence** none **Optical nature** isotropic **Dispersion** 0.057

Andradite ranges from green, yellow, yellow brown to black, with varietal names given to different colours. Iron and titanium are the main causes of colour, and the bright green variety demantoid is coloured by chromium. Andradite has the highest refractive index, producing very bright vitreous lustre, and the highest dispersion, making for lively gemstones. It is named after Brazilian mineralogist José Bonifácio de Andrada e Silva, who studied garnets from Norway. Andradite is commonly found in calcium-rich skarns such as metamorphosed impure limestones. It typically forms a solid solution series with grossular and uvarovite.

Demantoid, the magnificent emerald green variety, is one of the most valued of all garnets. The presence of chromium causes the bright green colour, while variable iron can give brown or yellowish tints. The name means diamond-like due to its adamantine lustre, high brilliance and high dispersion with more fire than diamond. This is best seen in lighter body colours, masked

▼ Faceted 3.25 ct demantoid exhibiting high dispersion, from Ural Mountains, Russia.

by darker colours, however for some the intense green is preferable.

This variety was first discovered in the mid 1800s in the Ural Mountains in Russia, and this remains the main source for high quality stones. Most faceted gemstones are less than a few carats, fine stones above 5 ct are very rare. Other sources include Italy, Switzerland, Iran and Madagascar. Demantoid is found in serpentinites and chlorite schists near ultramafic rocks, which provide the chromium. This gem was also discovered in Namibia in the 1990s, coloured by vanadium with only minor chromium.

Demantoid is normally transparent and often eye clean. It may contain 'horsetail' inclusions with clusters of fine hairlike chrysotile or tremolite-actinolite, often radiating from a chromite crystal. This inclusion is diagnostic of demantoid, typical of Russian stones (although found from other locations), and is even desirable in faceted gemstones. Other inclusions may be spinel or magnetite, liquid and two-phase inclusions. Demantoid is normally faceted in brilliant or cushion cuts to make the most of the dispersion and brilliance. It is one of the few garnets that may be treated, heated to remove yellow tints and improve the green colour. Demantoid is the softest of the

▼ A classic 'horsetail' inclusion with hairlike crystals radiating from a dark chromite inclusion. FOV 1.53 mm.

garnets and can be brittle, so extra care should be taken with this stone. Demantoid may be imitated by green glass or yttrium aluminium garnet (YAG).

Topazolite is pale to dark yellow, yellow-green or brown, and is generally only found as small crystals, but may be gem quality. It is named for its resemblance to topaz. It is found in metamorphic rocks, most notably from the USA, Italy and Switzerland. Melanite is a black, sometimes red, titanium-rich variety, which occurs in well-formed crystals. It is named from the Greek 'melanos'

meaning black. It occurs in alkaline igneous rocks. An attractive iridescent andradite from Mexico and Japan is marketed as rainbow garnet. This greenish or brownish garnet contains lamellar (thin-layered) structures that cause thin film interference and diffraction of light, creating vivid iridescent colours on the crystal faces and within the crystals. This is normally opaque to translucent and may be cut en cabochon or faceted, often in freeform shapes. Mineral specimens are also popular with collectors.

Grossular

Composition $Ca_3Al_2(SiO_4)_3$ calcium aluminium silicate **Crystal system** isometric **Hardness** 6.5–7.5 **Cleavage** none **Fracture** conchoidal **Lustre** vitreous **SG** 3.50–3.74 **RI** 1.730–1.760 **Birefringence** none **Optical nature** isotropic **Dispersion** 0.020 (hessonite 0.027)

The name grossular comes from the botanical name for gooseberry – Ribes grossularia – as the original specimens found in Russia were gooseberry green. Grossular forms a solid solution series with andradite and uvarovite, but may incorporate any gem garnet species, and its properties vary widely. With a pure end member composition it is colourless, but this is rare, and it is usually found in a range of colours. The presence of manganese causes yellow to orangey red or pink, iron imparts yellow to orange or red, and vanadium/chromium causes a rare bright green. Different colours are given varietal names. Grossular forms by the metamorphism of impure calcareous rocks.

Hessonite, also known as cinnamon stone, is the orange brown variety. The colour is due to manganese and iron, and may range through to reddish or peachy. Hessonite is commonly included with colourless rounded crystals and

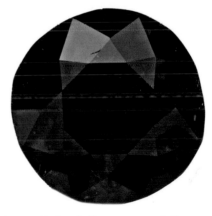

▼ A 10.84 ct mixed cut gemstone of grossular variety hessonite, also known as 'cinnamon' stone.

has a swirly, treacle-like appearance. As the eye visible inclusions affect the clarity, hessonite is not commonly used in jewellery today. It has, however, been used for cameo and intaglio engravings as far back as 2,000 years.

Hessonite is found in metamorphosed calcareous rocks and placer deposits. The best material comes from the gem gravels in Sri Lanka. Other sources include India, Madagascar, Brazil, Russia, the USA and the Jeffrey Mine in Canada, which produces popular collector specimens.

▲ A pear shaped 9.72 ct
tsavorite garnet.

Merelani mint garnet is a light bluish-green tone from Tanzania, found in the same mines that produce Tanzanite. It is similarly coloured by vanadium and chromium and can be classed as a pale tsavorite.

Transvaal jade is the misleading trade name for green grossular from South Africa. It occurs as a translucent to opaque compact rock, coloured green by chromium impurities, and containing small black inclusions of magnetite. It may also occur in grey (mixed with minor zoisite) or pink (coloured by manganese impurities). This gem is normally polished into cabochons or used for carving to simulate jade. It fluoresces orange-yellow under X-rays and may fluoresce pinkish orange under UV light, which can distinguish it from true jade.

A pink grossular garnet found in Mexico is known as raspberry garnet, however it tends to be opaque and is rarely used as a gemstone.

Tsavorite is an intense green variety first discovered in 1967 in Tanzania, then across the border in Kenya in 1970. It is named after the Tsavo National Park where it is mined. Tsavorite has since been found in Madagascar.

This variety is transparent and often eye clean, with the pastel to deep emerald green colour caused by vanadium and minor chromium. Tsavorite was brought to the gem market by the jewellers Tiffany & Co. in the early 1970s and changed the world of green gemstones. With as good a colour as emerald, but higher hardness, clarity and brilliance, and a lower price, it was cause for competition. Today it is one of the most expensive garnets.

Tsavorite is faceted in a range of shapes to best show the colour and retain weight. Most faceted gemstones are less than 3 ct, rare above 5 ct and world-class above 20 ct. Tsavorite is not normally treated. It may be imitated by green glass or yttrium aluminium garnet (YAG).

▼ A polished cabochon
of grossular garnet
known as 'Transvaal
Jade', from the Bushveld
complex, South Africa,
used as a jade simulant.

Uvarovite

Composition $Ca_3Cr_2(SiO_4)_3$ calcium chromium silicate **Crystal system** isometric **Hardness** 7,5 **Cleavage** none **Fracture** conchoidal **Lustre** vitreous **SG** 3.71–3.77 **RI** 1.840–1.870 **Birefringence** none **Optical nature** isotropic **Dispersion** 0.027

◀ Small bright green crystals of uvarovite garnet on matrix.

Uvarovite is an attractive bright green colour due to the essential chromium in its composition. This garnet was first found in Russia and is named after Count SS Uvarov. It is found in metamorphic skarns and serpentinites, formed through hydrothermal alternation of chromium-bearing rocks. The finest gem-quality crystals come from Russia but tend to be very small, only a few millimetres in size. Larger crystals up to 2 cm (¾ in) are found in Finland but are more opaque and often cracked, so not suitable for gemstones.

This is the only garnet that occurs in one colour, and its emerald green can rival tsavorite and demantoid. Sadly, due to the small crystal size, faceted stones are rare. It is more commonly seen as clusters of crystals on matrix set into jewellery.

Other notable garnet varieties

Rhodolite is a rose red to violet red variety, with an intermediate composition between pyrope and almandine. As it is approximately 70% pyrope, some consider it a variety of pyrope. It is commonly transparent and eye clean, and tends to be brighter than either end member. The most desirable colour is purplish red, first known from North Carolina, USA. Today sources include Tanzania, Kenya, Mozambique, Sri Lanka, Brazil and Myanmar.

Colour change garnet is known, exhibiting different colours in daylight and incandescent light. It occurs in garnets of different compositions, and the most well known is a mix of pyrope and spessartine with a colour change similar to alexandrite, caused by the presence of vanadium and/or chromium. Colour change garnets are known from Sri Lanka (bluish green in daylight to purplish red in incandescent light), Tanzania

(bluish-green to purple), and more recently from Rokily, Madagascar (bluish to purple change, also the first known blue garnets). Other colour change combinations exist such as purple to pink, and Brazil produces a purple to red colour change.

▼ A 1.88 ct rhodolite from Sri Lanka, resembling the purplish red of North Carolina rhodolite.

Mali garnet is dominantly grossular with andradite, and may be considered a variety of grossular. This occurs in bright yellow green to brownish green. It was discovered in 1994 in the Republic of Mali in West Africa.

Malaya (or malaia) garnet is a trade name used for garnets that do not fit into the garnet species or colour variety categories. There is no set definition of malaya garnet, but it is generally considered to be orange with pinkish, reddish, yellowish or brownish tints. The finest gemstones are pinkish orange without a brownish hue. It originally came from the Umba Valley on the Tanzania and Kenyan border, discovered in the mid-1960s and mined from alluvial deposits that had weathered from gneiss. This garnet is an intermediate spessartine-pyrope and some consider it a variety of pyrope as this is often the dominant species. Faceted stones are normally under 3 ct, although larger sizes were found early on. It is often eye clean, or with fine rutile needles. A more recent find of malaya garnet, in the late 1990s, occurred in Bekily, Madagascar, of pink to pinkish-orange garnets. These are intermediate pyrope-spessartine with

▲ Cushion mixed cut Malaya garnet of 29.29 ct, orange with yellow to brownish tints.

variable almandine and minor grossular. The colour is caused by variable iron and manganese. Other locations include Mozambique.

Mahenge garnet is the trade name for a pastel pinkish orange to purplish pink, found in Mahenge, Tanzania, around 2015. Many consider it malaya garnet. Similarly umbalite is a light pink to purple variety first found in the Umba Valley. This was not purple enough to be called rhodolite, however the trade name umbalite is not often in use, and some class this garnet as malaya with a purplish tint.

Iolite

Composition $Mg_2Al_4Si_5O_{18}$ magnesium aluminium silicate **Crystal system** orthorhombic **Hardness** 7–7.5 **Cleavage** good **Fracture** conchoidal, uneven **Lustre** vitreous **SG** 2.53–2.78 **RI** 1.522–1.578 **Birefringence** 0.005–0.018 **Optical nature** biaxial +/- **Dispersion** 0.017

▼ A polished slice of cordierite variety iolite from Finland.

Iolite is the violet-blue gem variety of the mineral cordierite, with colour to rival fine sapphire and tanzanite. It is best known for its strong trichroism, showing three different pleochroic colours as the gemstone is turned. The effect is so distinct that iolite is also known as dichroite. Cordierite was first studied by French geologist Pierre Louis Antoine Cordier, after whom it is named. The name iolite is from the Greek 'ios' meaning violet.

Part of the orthorhombic crystal system, iolite forms short prismatic crystals often appearing pseudohexagonal. Crystals are typically striated along the length. It may also occur as massive or granular material. Iolite is found in a range of environments, typically in metamorphic rocks associated with other gem minerals including sillimanite, spinel or garnet, as well as igneous and pegmatitic rocks. Most gem-quality material is found in alluvial placer deposits as waterworn pebbles, with Sri Lanka the main source. Iolite is also found in Myanmar, India, Madagascar, Tanzania, Kenya, Brazil, Norway, Russia, Canada and the USA. High-quality material is not found in reliable quantities to provide a steady supply, which restricts the popularity of this affordable gemstone.

Iolite is typically blue to violet from the presence of iron, although may be grey, yellowish brown or greenish. The strong pleochroism is easily seen. Blue stones exhibit deep violet-blue/pale bluish grey/pale straw yellow, and violet gemstones show dark violet/light violet/yellow brown pleochroism. The strong colour difference of deep blue to near colourless has led to the misleading trade name water sapphire.

It is transparent to translucent, and most faceted gemstones are eye clean to slightly included. Iolite from Sri Lanka may have tabular inclusions of iron oxides giving a reddish aventurescence, known as 'bloodshot' iolite. On occasion it may be chatoyant with many parallel tubular inclusions, producing a cat's eye effect when cut en cabochon, and very rarely asterism with a weak four-rayed star.

Iolite is faceted in a variety of cuts to highlight the transparency and colour. The pleochroism makes it challenging to cut. The deepest blue colour is seen down the length of the crystal. Rough is orientated to gain the finest colour when the gemstone is viewed face up, with the greatest

▲ A 19.69 ct fancy cut from Mogok, Myanmar with a central flaw. It is cut to show pale yellow and darker violet blue pleochroic colours face up.

▼ A 2.69 ct mixed cut iolite from Mogok, Myanmar.

weight. High-quality gemstones more than 5 ct are rare. Lesser quality material may be fashioned as cabochons or beads.

While relatively hard at 7–7.5 on Mohs scale, iolite can be brittle and has distinct cleavage in one direction. It should be treated with care and is best used in earring and pendant settings which offer more protection. Avoid knocks and sudden changes in temperature.

Iolite is not normally treated, which for many is a desirable feature. Synthetic cordierite is used in industry, but not normally seen on the gem market. Glass imitations of iolite are known, and it may be confused with sapphire, tanzanite and benitoite.

Jade

Jade has been revered for millennia for its colour, beauty and, most of all, its toughness. It has been used since the Stone Age, shaped into axe heads and other tools, because of its ability to withstand fracture and retain a sharp edge. The Olmecs, Mayans and Aztecs of Mesoamerica valued jade over gold and it was used for jewellery, ornaments, ceremonial items and medicinal purposes. The Māori of New Zealand used jade in all aspects of life, from tools and weapons to treasured possessions. But it is most prized in China, valued above all other gem materials. Here, its history goes back at least 7,000 years, and the Chinese character 玉 for yù (jade) is said to be one of the earliest written.

Jade is considered a stone of heaven, representing virtue, bringing peace, purity and nobility, and giving resilience and protection to those who wear it. It has been consistently intertwined with Chinese culture, and the artform of carving jade continues today. While jade was used in Europe for tools and ritual items in the Neolithic period, around 4,000 to 8,000 years ago, it was not appreciated as a precious gem until the Spanish brought it back from the Aztecs in the 1500s. Today, jade is the second most valuable gem (after diamond) in the global jewellery industry, with the finest quality commanding higher prices per carat than the best rubies or sapphires, beaten only by rare fancy colour diamonds.

The name jade is generally agreed to originate with the Spanish explorers of Mesoamerica in the 1500s, derived from the Spanish 'piedra de ijada/hijada', meaning stone of the flank or stone of pain in the side for the reputed ability of jade to cure kidney problems when worn. This was corrupted through the French 'pierre le jade' to the English jade.

Jade is not a single mineral species like many other gems. The term encompasses two different metamorphic rocks. One is composed of pyroxene minerals, primarily jadeite, and is known as jadeite jade. The other rock, called nephrite jade, consists of amphibole minerals, predominantly tremolite-

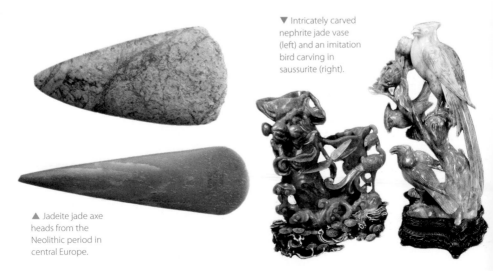

▼ Intricately carved nephrite jade vase (left) and an imitation bird carving in saussurite (right).

▲ Jadeite jade axe heads from the Neolithic period in central Europe.

actinolite. Both rocks are compact aggregates of tiny interlocking crystals. This is what gives jade its extreme toughness, allowing it to be intricately carved without fear of breaking or chipping. Although previously recognised in China that the green jade from Myanmar was different to the jade from China, it was not until 1863 that French mineralogist Augustin Alexis Damour proved this by analysis, and proposed the name jadeite for the green jade to distinguish it. The name jadeite was derived from jade, and nephrite originates from the Latin 'lapis nephriticus' for kidney stone, in turn derived from the Ancient Greek 'nephrós' for kidney.

The identification of jade is not simple, but critical due to its value. Not only is it composed of two different rocks, but the composition of each rock can vary and is not homogeneous. A combination of visual assessment, specific gravity, hardness, refractive index and absorption spectrums may distinguish between jadeite and nephrite jade, and more importantly simulants. However, analysis by gem laboratories using equipment such as Raman spectroscopy and X-ray fluorescence is often required to determine the crystal structure and chemical composition. The identification of any treatments, particularly of jadeite jade, is of similar importance.

As a highly valued gem, there are unsurprisingly many jade imitations.

Green or mottled rocks include:

- saussurite (a dark green and white rock composed of zoisite, feldspar and other minerals, and used for carving)
- dyed quartzite (distinguished by dye between interlocking quartz grains)
- maw-sit sit (a bright to dark green rock with black spots and green veins, composed of many minerals including chromium-rich kosmochlor, albite, other amphiboles and chromite), named after the village in Myanmar near where it was found

Mineral simulants, and possibilities for confusion, are:

- bowenite (variety of serpentine mineral antigorite)
- serpentine (a mottled mixture of very soft, less dense serpentine group minerals, with a greasy feel)
- chrysoprase (bright green nickel-rich variety of chalcedony, with uniform colour and conchoidal fracture)
- aventurine quartz (with tiny included green platelets of chromium-rich mica)
- hydrogrossular or Transvaal jade (variety of grossular garnet)
- californite or American jade (a green cryptocrystalline variety of vesuvianite with a similar appearance to jade)
- feldspar variety amazonite (distinguished by white streaks, incipient cleavage, and lower lustre)

Human-made imitations include:

- glass and plastic (distinguished by bubble and swirl inclusions, is warm to the touch and glass may show conchoidal fracture)

Composite triplets are also known, of a translucent jadeite base and cap sandwiched around a green glue or cement.

▶ A polished bead of maw-sit-sit from Maw-sit-sit, Myanmar, used to simulate jade.

Jadeite jade

Composition NaAlSi$_2$O$_6$ sodium aluminium silicate **Crystal system** monoclinic (microcrystalline) **Hardness** 6.5–7 **Cleavage** none **Fracture** splintery, uneven **Lustre** vitreous, greasy **SG** 3.25–3.50 **RI** 1.640–1.688 **Birefringence** 0.012–0.020 **Optical nature** biaxial + **Dispersion** none

▼ Jadeite jade boulder, 22 cm (8.5 in) with brown weathered skin into which 'windows' are polished to reveal the inner colour.

Jadeite jade is a type of jade, and one of the most treasured gemstones in the world, particularly in Asia. It is rarer, more translucent, slightly harder, nearly as tough, and more valuable than the other type of jade, nephrite. Jadeite jade is a metamorphic rock composed almost entirely of the pyroxene mineral (confusingly with the same name) jadeite. Other pyroxenes such as omphacite and kosmochlor (NaCrSi$_2$O$_6$) may be present, and occasionally these occur as the dominant mineral in jade. The Chinese term 'fei cui' (after the kingfisher bird for its bright green feathers) is also officially used for jade composed of jadeite, omphacite and kosmochlor in any combination.

Jadeite jade is found in few locations around the world. It was prized by the Olmecs, Mayans and Aztecs in Mesoamerica, and Guatemala is still an important source. The finest jadeite, especially vivid green, comes from Kachin State in northern Myanmar, and this is the largest and most important source. It is also found in Russia, Kazakhstan, the USA (California), Japan, Italy, Dominican Republic, and Cuba. High demand in China in the mid- to late-1700s for fine green jadeite from Myanmar created a jade trade that still flourishes. As one of the main cutting centres, and also the largest consumer, most imported jadeite jade is marketed and sold in China without entering the international market.

The mineral jadeite requires specific conditions to form, occurring in high pressure and relatively low temperature metamorphic environments.

It is produced by the metasomatic alteration of sodium-rich parent rocks from fluids in subduction zone settings, forming nodules or lenses hosted in, or adjacent to, serpentinite. Jadeite crystals occur as tiny interlocking grains and only very rarely as small bladed crystals. Some deposits are mined in situ from serpentinite, however due to jadeite jade's ability to withstand weathering, most is mined from alluvial deposits as boulders to pebbles with the primary source not always known. The boulders have an outer weathered rind or skin of variable thickness. The rough jade is often sold with a 'window' polished through the skin to reveal the inner colour, which can vary. Buyers determine the quality of the whole boulder by shining a torch through the window. They judge the texture and transparency by how the light penetrates the stone, and the colour by how it continues into the stone. Individual pieces of rough may be hundreds of kilogrammes (thousands of pounds) in weight and top quality rough sells for millions of US dollars.

Jadeite jade is most widely known in green, but can occur in all colours or attractively mottled. Pure jadeite is white, but the presence of iron creates yellowish through to bluish greens, and chromium imparts intense emerald greens. Lavender colours are caused by manganese impurities, and black may be due to graphite micro-inclusions. Iron oxides along grain boundaries can give brownish colours, which produce orange or red colours through heat treatment.

Jadeite jade is semi-transparent to opaque and has a grainy appearance due to the tiny interlocking crystals. As the grains have different orientations, they have variable hardness, which leads to undercutting of some grains during polishing giving a pitted 'orange peel' surface. This can distinguish jadeite jade from nephrite. The lustre when polished ranges from greasy to vitreous, and tends to be sharper than that of nephrite.

The crystal size determines whether the texture of jadeite jade is fine, medium or coarse. The smaller the grain, the finer the texture, and this creates a higher quality jade with a smoother surface, brighter lustre and higher transparency. Typical inclusions are black, brown or dark green minerals, white spots, and both healed and unhealed fractures. Feathers are commonly seen under magnification.

The value of jadeite jade is determined by the colour, colour intensity, transparency and texture. The most prized jadeite is a near transparent vivid green known as imperial jade. Lavender is the next valued, especially when the colour is saturated. Black, red and orange are gaining popularity. Olmec blue is a rare blue-green from Guatemala. With all colours, the ideal stone is semi-transparent with an intense even hue with no spotting, mottling or veining. It should have a fine texture with no inclusions, allowing light to pass through the stone so the colour glows from within. Semi-transparent material is still valued even if the colour is pale or uneven. Jadeite jade with contrasting mottling or veining, which is often considered less valuable, can be very attractive.

High-quality jadeite jade is typically cut as moderately domed cabochons to highlight the colour and translucency. It is also fashioned in round or carved beads, and strands can fetch high prices due to the difficulty in matching colour, transparency and texture of the individual beads. Jewellery may be fashioned out of single

▼ Translucent jadeite jade cabochons of imperial jade (6.60 ct) and lavender (5.90 ct) colour, both from Myanmar.

pieces of rough, including rings, pendants and the very popular bangle bracelets. Bangles have high spiritual value believed to bring protection to the wearer, which combined with the large amount of rough jadeite required means they command some of the highest prices. As well as jewellery, jadeite is carved into statues, vases, bowls and many other items for decorative and religious use. Mottled colours or streaks, as well as the weathered brown skin, may be cleverly incorporated into the design. Carvings are usually intricate in a range of sizes, and the craftsmanship is valued as much as the jadeite itself.

◀ A carved pendant of jadeite jade.

Jadeite is often treated. Its granular texture is porous allowing it to be enhanced by dyes and impregnation. The most common treatments are given simple gradings.

A jade (or Type A) is natural untreated jade, or with a superficial colourless wax treatment to improve the lustre.

B jade has been bleached with strong acids to remove brownish colours caused by iron staining, and dark inclusions. This weakens the jadeite jade, making it brittle, so it is stabilized by impregnation with a polymer of similar refractive index. This fills any voids and cracks, improving the clarity, lustre and durability. This is a common treatment and can be difficult to detect. Polymers may give an evenly distributed fluorescence under longwave UV light. Polymers may degrade over time, becoming cloudy, and the jadeite can still be brittle so should be treated with care.

C jade is jadeite jade dyed green, lavender or other colours. It may be detected under magnification as concentrations of colour along grain boundaries and fractures. Dyed jadeite may appear red under the Gem-A Chelsea Colour Filter (due to chromium in the dye), while undyed jadeite remains green. Dyes may also react under longwave ultraviolet light while untreated jadeite is generally inert. Lavender dyes can give an orange response. This treatment is not stable and the colour may fade.

B+C jade is bleached then dyed and stabilized.

Heat treatment of jadeite jade containing iron oxide inclusions may produce a stable orange or reddish colour, and this treatment is undetectable. All treatments of bleaching, impregnation, dyeing and heating should be disclosed.

Synthetic jadeite has been produced in green and lavender, initially by the American conglomerate General Electric in the late 1970s, but is not seen commercially on the market.

Nephrite jade

Composition $Ca_2(Mg,Fe)_5(Si_4O_{11})_2(OH)_2$ calcium magnesium iron silicate **Crystal system** monoclinic (microcrystalline) **Hardness** 6–6.5 **Cleavage** none **Fracture** splintery **Lustre** greasy, vitreous **SG** 2.90–3.15 **RI** 1.600–1.641 **Birefringence** 0.027 **Optical nature** biaxial - **Dispersion** none

▶ Hei-tiki carved in nephrite jade with inlaid eyes of pāua shell, from New Zealand.

Nephrite jade has been used since at least the Neolithic period, as axe heads in Britain and in Europe. It has a similarly long history in China, New Zealand and North America, used for tools, adornment and ritual objects, and maintains cultural significance. The Māori in New Zealand are well known for their tools, and neck ornaments known as hei-tiki, carved in fine dark green nephrite, also called greenstone. The ancient

Chinese placed nephrite amulets in the mouths of their deceased, even burying royalty in jade burial suits. Nephrite is laborious and slow to fashion due to its remarkable toughness, yet the craftmanship of jade carving has been developed in China since long before the introduction of metal tools.

All ancient jade in China was nephrite, as it was not until the 1700s that jadeite jade was imported from Myanmar. The most important sources were the alluvial deposits of the White Jade and Black Jade Rivers near Khotan (now Hotan or Hetian, Xinjiang Province) and their primary source in the Kunlun Mountains. Although many historical sources are depleted, important deposits still exist in northwestern China, and the name Hetian jade is synonymous with the finest quality nephrite. Today the main sources are British Columbia in Canada, Siberia in Russia, South Korea, the USA (Alaska, Wyoming and California), New Zealand, Australia and Poland. Nephrite is more widely available than jadeite jade and more affordable.

Nephrite is a metamorphic rock composed of amphibole minerals, predominantly tremolite with variable actinolite. Tremolite is a calcium magnesium silicate $Ca_2Mg_5(Si_4O_{11})_2(OH)_2$ which forms a solid solution series with actinolite. Iron substitutes for magnesium in the crystal structure, and when it reaches over 10% it is defined as actinolite. Pure tremolite is creamy white, and as the iron content increases towards actinolite, the nephrite becomes greener and darker. Microscopic inclusions of other minerals and impurities cause a variety of other colours. The crystals form a dense microcrystalline aggregate with a fibrous 'felted' texture making nephrite tougher than jadeite jade and, in fact, the toughest of all gems. The smaller the microcrystals, the closer to tremolite composition (white) and the purer (without other mineral impurities), the higher the gem-quality of the nephrite jade.

◀ A carved pendant of white 'mutton fat' nephrite jade.

This type of jade forms through metamorphism and metasomatism of magnesium-rich rocks under low to medium pressure and temperature conditions. It occurs associated with serpentinite rocks in ophiolites, which are rocks of the oceanic crust that have been uplifted to the surface by the collision of oceanic and continental tectonic plates, obducted over the top of the continental crust. Many nephrite deposits are therefore in countries around the Pacific Rim. Less frequently nephrite occurs at the contact zone between dolomitic (magnesium) marbles and magmatic rocks. As nephrite is so tough, it is weathered and carried far from the source, and most is mined as pebbles to boulders in river or beach alluvial deposits. The boulders may reach several tonnes (several thousands of pounds) in weight. These often have weathered skins of yellow, orange or brown iron oxides.

Nephrite occurs in a range of colours through white, green, yellow, brown and black. The colours are less vivid than jadeite and usually mottled. Translucent white or cream-coloured nephrite (pure tremolite) is known as mutton fat jade and is the most valued, famous from the Hetian area. Green is next, although generally darker and more yellowish than jadeite jade. Spinach green nephrite, often with black dots of chromite inclusions, is considered the finest green colour, coming from the Sayan Mountains in Siberia

▶ A delicate paper-thin nephrite jade bowl, 12 cm (4.7 in) wide.

and also British Columbia in Canada. The brighter green colour is caused by chromium impurities. Yellow and brown colours are caused by iron oxides, and grey to black by microscopic black mineral inclusions such as graphite. Nephrite may exhibit a cat's eye effect when cut en cabochon. This occurs due to alignment of the amphibole fibres causing chatoyancy, and is known from Siberia, China and the USA.

The fibrous structure of nephrite is too small to detect under magnification, therefore it always has a fine texture and smooth surface, never showing the 'orange peel' surface texture sometimes seen on jadeite jade. The lustre is greasy, and softer than jadeite due to the slightly lower hardness. Typical inclusions are tiny grains of other minerals including magnetite, chromite, graphite, diopside and garnet.

Nephrite tends to be less translucent than jadeite jade. As such it is more commonly used for delicately carved jewellery than cabochons. The variable colours and weathered brown rinds are skilfully incorporated into designs in cameos. It is carved into ornaments such as vases and statues, which may be elaborate or eggshell thin. Interestingly, nephrite also has musical properties and rings like a bell when struck. Chimes and gongs have been created from nephrite in different sizes to play certain musical notes.

Nephrite is less commonly treated than jadeite jade. Due to its dense fibrous structure it is less porous, so is rarely dyed. It may be surface treated with wax, impregnated, or heat treated to improve the colour or create an antique weathered look, adding value.

Kyanite

Composition Al_2SiO_5 aluminium silicate **Crystal system** triclinic **Hardness** 4–7 **Cleavage** perfect **Fracture** splintery **Lustre** vitreous **SG** 3.53–3.75 **RI** 1.709–1.735 **Birefringence** 0.012–0.021 **Optical nature** biaxial - **Dispersion** 0.020

Kyanite forms distinctive blue, bladed crystals and creates magnificent faceted gemstones of intense blue. The name kyanite comes from the Greek 'cyanos' meaning deep blue, and was commonly spelt cyanite until the early 1900s. It is most well known for its extremely variable hardness in different directions of the crystal – lengthways the hardness is 6–7, and crosswise 4–4.5, hence it was also known as disthene from the Greek for double and strength.

As a common metamorphic mineral, it occurs in gneiss and schist, but may also be found in granitic pegmatites and secondary alluvial deposits. It is the high pressure polymorph of Al_2O_5, having the same composition as andalusite and sillimanite, but different crystal structure. Kyanite has triclinic crystal symmetry, forming as long bladed or flattened tabular crystals, and is a popular collector mineral due to the lovely

◀ Blue bladed crystals of kyanite with white quartz from Minas Gerais, Brazil.

green crystals showing three shades of green. The finest gemstones are sapphire blue with violet pleochroic flashes. The blue colours are caused by impurities of iron and titanium, green is from vanadium and orange from manganese.

Cutting kyanite is challenging due to the differential hardness, bladed crystal shape and perfect cleavage. Faceted gemstones are mostly elongate step cuts determined by the long crystal shape. Gemstones are small but may reach up to 15 ct. Lesser material is polished into cabochons. Chatoyant stones exhibiting cat's eye effect and asterism do occur on rare occasions, and even a change colour effect is known. Kyanite is not usually treated. Care should be taken when wearing due to the lower hardness and perfect cleavage. It may be confused with sapphire, tanzanite and iolite due to their similar colours.

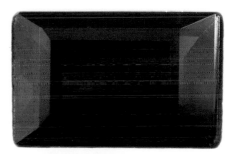

▲ Kyanite step cut 2.53 ct gemstone from Myanmar.

◀ Elongate step cut 2.74 ct kyanite gemstone from Loliondo, Tanzania, of rare orange colour.

colour and distinctive shape. It is not common in gem quality, and the lack of consistent colour and transparency in the crystals limits the yield. Brazil is the main producer with locations in Minas Gerais and Bahia. Other sources include Cambodia, Myanmar, Russia, the USA, Kenya, Madagascar, Mozambique and India. Exceptional sapphire blue crystals have been found in Nepal since 1995.

Kyanite is normally blue, ranging from saturated deep blue to pale blues, or greenish blue. Crystals commonly contain a range of shades, and specimens with a deep blue central stripe down the crystal length are well known from Vitória da Conquista, Brazil. Occasionally kyanite may be green or teal (found in Umba Valley, Kenya) and rarely orange (first found in garnet-bearing mica schists in Loliondo, Tanzania in 2008). Kyanite is strongly trichroic with blue crystals exhibiting dark blue/violet/pale blue to colourless, and

Lapis lazuli

Composition mixture predominantly lazurite, calcite, pyrite **Crystal system** n/a **Hardness** 5–6 **Cleavage** poor to none **Fracture** uneven **Lustre** waxy, vitreous **SG** 2.38–3.00 **RI** 1.500–1.550 **Birefringence** none **Optical nature** isotropic **Dispersion** none

▼ A polished slice of lapis lazuli.

The popular lapis lazuli is not actually a mineral, but a rock typically composed of deep blue lazurite, white calcite veining and gold flecks of pyrite. The name means blue stone, and it may simply be known as lapis. It has been used since ancient times as a decorative stone, prized for its rich colour. There is evidence that deposits near Sar-e-Sang, in Badakhshan Province, Afghanistan, were mined thousands of years ago, possibly as early as 7,000 BC. The ancient Egyptians treasured lapis lazuli, using it as an inlay in the gold burial mask of King Tutankhamun. Lapis lazuli was ground to produce the ultramarine pigment, the finest and most expensive blue pigment for centuries until its replacement by a synthetic variety in the early 1800s. Afghanistan is still the main source. Russia and Chile are significant sources, and minor quantities come from Italy, Myanmar, Canada (paler), the USA (darker), Argentina and Pakistan.

Lapis lazuli is a metamorphic rock that occurs in marble. The most important mineral component is lazurite, a member of the sodalite group. Haüyne and sodalite, other blue minerals of the sodalite group, may also be present, and some scientists consider lazurite a sulphur-rich variety of haüyne. These minerals give lapis lazuli its intense ultramarine colour, caused by the presence of sulphur. The finest quality lapis lazuli has a uniform intense blue to violet-blue colour with accents of pyrite. More commonly it is paler to darker blue or greenish blue, patchy with variable white calcite and small metallic inclusions of pyrite. It is very popular in jewellery as polished cabochons or beads, which highlight the colour and markings. It is carved into ornate objects or utensils, and polished slices are used for inlays and decorative stone. Lapis lazuli may occur in very large blocks, so carved objects can be sizeable.

As lapis lazuli is an aggregate of minerals, its properties change with composition. It is normally opaque and the lustre ranges from waxy to vitreous. Cleavage is poor. Its specific gravity varies, becoming higher with increased pyrite content. It is relatively soft at 5–6 (individually lazurite is 5-5.5 and calcite 3, but other minerals increase the overall hardness), so may be scratched and will crack or chip under pressure. Care should be taken when wearing, and protective settings are recommended for rings. It is sensitive to heat and susceptible to

▶ Necklace of polished lapis lazuli beads.

▼ Polished beads from Afghanistan, in order of increasing quality (left to right).

acid attack from household chemicals, hairspray and perfume. Some lapis lazuli may be dyed to improve the blue colour or hide white calcite veining. It may also be impregnated to improve the surface lustre and hide cracks.

Imitation lapis lazuli is made by Pierre Gilson in France, known as Gilson lapis. It is of similar composition containing lazurite, but softer. It may be distinguished by lower specific gravity, lack of calcite and does not have the random appearance of natural material. Other simulants include the similarly coloured sodalite, dyed jasper, dyed howlite, dyed magnesite, and a grainy synthetic spinel coloured by cobalt with thin inclusions of gold.

▼ A carved bowl in lapis lazuli with Egyptian motif.

Malachite

Composition $Cu_2(CO_3)(OH)_2$ copper carbonate **Crystal system** monoclinic **Hardness** 3.5–4 **Cleavage** perfect **Fracture** splintery, uneven **Lustre** vitreous, silky **SG** 3.25–4.10 **RI** 1.655–1.909 **Birefringence** 0.250–0.254 **Optical nature** biaxial - **Dispersion** none

Malachite is a copper carbonate mineral with vivid colour caused by the copper, and characteristic banding in a range of greens. The name is thought to be derived from its resemblance to the mallow plant, and it has been known as malachite since the latter half of the 1600s.

Known since antiquity, malachite is mentioned by Roman philosopher Pliny the Elder, and there is evidence of mining it for copper as far back as the Bronze Age in Wales. It was favoured by the ancient Egyptians, Greeks and Romans for use in amulets and jewellery, even powdered for eye shadow. Malachite was used as a mineral pigment for its bright green colour until about 1800, before being replaced by synthetic pigments. Today it is still mined as a minor ore of copper.

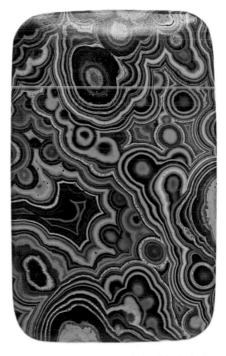

▲ A polished cabochon revealing the attractive light and dark banding of malachite.

▶ A magnificent banded malachite, 21 cm (8 in) wide from Zambia.

Malachite is a common mineral, found as an alteration product in the oxidation zones of copper deposits. It is frequently found in copper deposits near limestone, which provides the carbonate. It is more prevalent than the blue copper carbonate mineral azurite, with which it is often found. The Ural Mountains, Russia, was historically one of the major sources. Today the main producer is the Democratic Republic of the Congo. Other sources include Tsumeb in Namibia, Burra Burra in South Australia, Peru and southwest USA particularly Arizona.

Part of the monoclinic crystal system of symmetry, malachite is rarely found as acicular to prismatic primary crystals. It normally occurs in fibrous or velvety layers, or crystalline aggregates in botryoidal clusters and stalactitic forms. Due to its close relationship with azurite, they are commonly found intergrown, known as azurmalachite, or as azurite altering to the more stable malachite, forming pseudomorphs retaining the prismatic crystal shape of the azurite. Less commonly malachite replaces other copper-bearing minerals such as cuprite.

Normally opaque, malachite's colour ranges through pale, bright or deep green to nearly black. As rough material it has a vitreous or earthy lustre. It takes a good polish, appearing silky with a chatoyant effect when fibrous. When sliced and polished the alternating pale and dark green banding is quite beautiful. Malachite is understandably a desirable decorative stone, carved or used as inlay for tables, columns and object d'art. Often slices of malachite are fitted together like a mosaic and used to create a thin veneer. It has been used extensively worldwide, highly prized by the Russian royal family in the 1800s, as seen in the famed Malachite Room in the Hermitage Museum in St Petersburg. In jewellery, malachite is fashioned into cabochons and beads to highlight the attractive banding, with patterns of concentric rings very popular.

Similar to azurite, malachite should be treated with care. It is soft and easily scratched, registering 3.5–4 on the Mohs scale of hardness. It is sensitive to light and reacts with acids, even carbonated water. As with azurite, its dust is toxic and must not be inhaled when cutting and polishing. Malachite may be coated with wax or the surface impregnated with resin to improve the hardness and lustre.

Opal

Composition $SiO_2.nH_2O$ hydrated silica **Crystal system** amorphous **Hardness** 5–6.5 **Cleavage** none **Fracture** conchoidal **Lustre** vitreous **SG** 1.25–2.50 **RI** 1.370–1.520 **Birefringence** none **Optical nature** isotropic **Dispersion** none

Opal is perhaps the most beautiful gemstone, with bright colourful flashes of iridescence that appear to float magically across it. Known as play of colour, this optical effect shows spectral colours appearing to move as the stone is turned. The hue and size of the flashes are unique to each opal and can be every colour of the rainbow.

It is a fascinating gem: its naming and history is somewhat mysterious, it is considered both lucky and cursed, and the mode of formation and cause of spectral colour has puzzled scientists for years. The name is thought to come from the Sanskrit 'upala', Greek 'opallios' or Latin 'opalus' meaning stone or precious stone. The description of 'opalus' by Roman philosopher Pliny the Elder in 77 AD accurately describes the play of colour, but few opals are found in ancient artefacts. Have they not survived, or was it not the opal of today? Opal lore has changed considerably from a symbol of love, hope, magic and luck, to one of a curse, with links to the supernatural. Interestingly this change in perception is attributed to the novel *Anne of Geierstein* written in 1829 by Sir Walter Scott, but the superstition persists.

Mineralogically, this enigmatic gem is simply hydrated silica with a water content varying up to 20%, but normally 6–10%. Opal is a secondary mineral, forming after the host rock, and not found at depth. The exact formation mechanism is still unknown, but it is generally agreed that opal forms from a silica-rich fluid permeating into cracks and voids within Earth's crust, slowly concentrating into a viscous gel then hardening into solid opal. It occurs as thin seams, veins and nodules. Opal may also replace other structures such as fossil animals

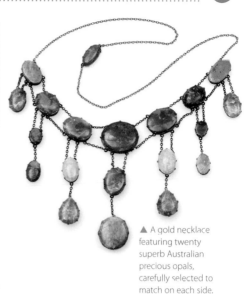

▲ A gold necklace featuring twenty superb Australian precious opals, carefully selected to match on each side.

or plants, known as pseudomorphs. One of the most famous is the opalised pliosaur named Eric found in Coober Pedy in South Australia, now part of the Australian Museum collection.

While traditionally considered a mineral, opal is more correctly classed as a mineraloid as it is amorphous without a crystal structure. Instead, it is made up of tiny spheres of silica, averaging 150 to 300 nm in diameter, arranged in a regular 3D array. Its structure was only discovered in 1964 when opal was studied using a scanning electron microscope. The mystery of the play of colour phenomenon was solved when the structure of opal was understood. The silica spheres in their regular array act like a diffraction grating. When light passes between the spheres it is diffracted and spread out. The light waves interfere with others, boosting some colours and cancelling others out. The size of the spheres, and therefore the gaps, are smaller than the wavelengths of visible light (~ 400 to 700 nm), and determine the colours seen. Smaller spheres diffract smaller wavelengths, resulting in violets, blues and greens. Larger spheres diffract longer wavelengths up to reds, and opals with a mix of sizes potentially diffract all wavelengths with a rainbow of colours. Precious opal has regions of spheres of uniform size in orderly rows. Individual regions are slightly varied in their orientation and sphere size. This means each region diffracts the light differently, creating an individual colour flash. Together they show a range of spectral colours, changing as the stone is turned. The more regions, the more lively the stone.

Another optical effect associated with opal is opalescence. This term is used to describe a pearly or milky shimmer seen in opal and other minerals, and is a form of adularescence. Opalescence is caused by the scattering of light and is different to play of colour. Many opals fluoresce under longwave and shortwave UV light with a white, bluish, greenish or brownish response. Many also show phosphorescence, continuing to emit light after the UV light source is removed.

Opal has a background or body colour determined by the composition of its inclusions and/or impurities such as iron or nickel. It may be white, orange, red, brown, yellow, green or black. This is different to the play of colour caused by its structure. Most opal is translucent but may be opaque or transparent.

1 μm

▲ The internal structure of opal, consisting of tiny silica spheres in a regular array, revealed by a scanning electron microscope.

This gemstone is divided into three main groups: precious opal (or noble opal) showing a play of colour, common opal (or potch) which does not show a play of colour, and fire opal of bright orange body colour with, or without, a play of colour. Each group has a multitude of varietal and trade names based on appearance and source. The main types are described below.

Precious opal

Varieties are defined by their body colour and play of colour. The combination of colours, and size of flashes determine its value. Precious opal was rare until the 1800s when 'Hungarian' opal was commercially produced from deposits in modern-day Slovakia. These are the most significant deposits in Europe, with mining known since at least the 1500s but possibly as far back as Roman times. The discovery of opal in Australia around 1840, followed by commercial production from 1875, brought stiff competition and Australia has been the world's main producer of precious opal for the past century. Here the opal is found in old weathered sedimentary rocks, different from many other sources where it formed in volcanic rocks. Mexico has a long history of opal mining with spectacular varieties and was for some time the second largest producer. Ethiopia is now a significant supplier, with opal first found in the early 1990s. Other sources are the USA, Canada, Brazil and Honduras.

- **White opal** – the most common variety of precious opal, with a translucent to semi-translucent light body colour of white, yellow and grey. The main source is Coober Pedy in South Australia. Opal was first found here in 1915, in the largest opal field in the world. Other sources include Andamooka, Mintabie and White Cliffs in Australia, Slovakia, Ethiopia, Canada, Brazil and Honduras. Ethiopian white opal was discovered in 2008 in the Wollo province, known as welo opal. Many are hydrophane, meaning they readily absorb water, increasing their transparency and weight, at the expense of their play of colour. Fortunately, they dry out returning to their original weight and appearance, but it is advised to keep welo opals away from water.

- **Black opal** – the rarest and most valued precious opal, with a translucent to opaque dark body colour of blue, grey or black. The dark backdrop enhances the spectral play of colour. Lightning Ridge in New South Wales, Australia is the main source, where it has been mined since 1901. Many pseudomorphs of fossils occur here. Andamooka and Mintabie in South Australia produce small quantities, as do Slovakia, Brazil, Honduras, Indonesia, the USA and Mexico.

- **Water or crystal opal** – with colourless transparent to semi-transparent body colour, in which the play of colour magically floats. The best source is Mexico, but it is also found in Australia, the USA, Ethiopia and Brazil.

- **Boulder opal** – occurring as thin veins and seams of opal in an ironstone host rock. The opal is translucent to opaque with a beautiful play of colour on a light or dark body colour. The fragile layer is routinely cut and polished still attached to the brown matrix for reinforcement, and it is often incorporated into the design of the gemstone. Boulder opal was first discovered in Queensland, Australia, and mined from the 1890s. It was later found in Brazil and Canada, but is rare on the market from these countries. The opal may be in thin layers, a fine network of veins known as matrix opal, or a nodule inside a small concretion known as Yowah nuts when found near Yowah, Queensland, Australia.

- **Harlequin opal** – a precious opal where the play of colour is patterned in a mosaic or chess board arrangement. Originally applied to 'Hungarian opal', it now includes Australian.

▼ A 10.82 ct cabochon of black opal with superb play of colour. From Lightning Ridge, NSW, Australia.

▲ Boulder opal from Queensland, Australia on ironstone matrix.

Common opal

This occurs in a range of body colours but does not have a play of colour. The silica spheres are not of uniform size or orderly arrangement, therefore are unable to cause diffraction and interference of light. Common opal is typically opaque to translucent. It is frequently brownish orange, but ranges white, pink, blue, green, yellow, and even transparent colourless. Common opal is, perhaps unsurprisingly, common and found in many locations around the world. Peru is well known for producing a range of colours.

Common opal varieties:

- **Agate opal** – a banded variety of common opal, similar to agate or onyx in appearance. It may occur in veins or opalised fossils. The name however is misleading.

- **Andean opal** – the trade name for translucent blue to green opal found in Peru, coloured by microscopic inclusions of copper-rich minerals such as chrysocolla.

- **Moss or dendritic opal** – a translucent opal with moss- or branch-like inclusions of iron or manganese oxides or chlorite. Sources include the USA, Brazil and Australia among many others.

- **Pink opal** – a rare, mostly opaque pink opal from Peru, coloured by organic inclusions. It resembles pink coral but is of higher hardness.

- **Prase opal** – a translucent to opaque bright green opal coloured by microscopic inclusions of nickel-rich minerals. Originally sourced from Poland, it is also found in Tanzania and the USA.

- **Wood opal** – a petrified wood that has been replaced by opaline silica, often retaining the wood structure and concentric growth. It tends to be brown to cream in colour. It is well known from Australia and the USA.

- **Hyalite** – typically colourless and water clear. Considered the purest opal, it often has a botryoidal form resembling blobs of jelly or water. It commonly fluoresces under UV light. In 2013 a new find of hyalite opal in Zacatecas, Mexico, produced material with fluorescence so strong it caused a green to yellow body colour in daylight. The fluorescence under UV light is spectacular, caused by very low levels of uranium impurities.

Fire opal is transparent to translucent, sometimes opaque, with a vivid yellow, orange or red body colour. The colour is caused by microscopic inclusions of iron oxides. Some, but not all, fire opal has a play of colour, resulting in stunning

▲ Hyalite opal natural specimen and a 1.55 ct faceted gemstone from Zacatecas, Mexico, exhibiting strong greenish yellow fluorescence under UV light.

▼ 5.76 ct faceted fire opal from Mexico.

gemstones. Fire opal is primarily from Mexico, but found in other locations including Hungary, Turkey, Brazil, Canada and the USA.

Opal is soft with a hardness of 5–6.5 and brittle. Precious opal is cut to best show a vivid play of colour. Normally polished into cabochons, those with a well-rounded dome give intense results seen from all angles, and are the most valuable. Freeform shapes allow better use of the rough to show more play of colour and may incorporate the matrix into the design for additional support. Transparent to translucent opal is occasionally faceted, and those with a play of colour have spectacular results. Precious, fire and boulder opal are all used in jewellery. Pendants and earrings are common, offering a more protective setting. Common opal is often carved for ornamental use.

Most precious opal occurs in thin fragile layers, so to create gemstones it is assembled into composite stones. Known as doublets or triplets, these improve the strength and durability, and allow the opal to be useable as a gem. They increase the availability of opal on the market and have a more affordable price than solid opal. Examining the side of the gemstone under magnification is a simple way to spot the composite layers and joins. Doublets are formed of two layers, typically a thin layer of precious opal more than 2 mm (0.079 in) thick, backed by common opal, onyx, glass or plastic. A dark opaque backing will create the appearance of black opal. Slices of ironstone matrix may be used to imitate boulder opal. Opal that is cut and left attached to its protective matrix forms a natural doublet. The matrix may provide a light or dark background. These gemstones lack the distinctive join, with a natural irregular boundary between the opal and matrix. Boulder opal is typically cut this way to include the ironstone matrix as the backing. Similarly, if precious opal forms in a layer over common opal, it will be cut to use the common opal as the backing. Natural and assembled doublets can be capped by a transparent layer to form a triplet. The topping is typically rock crystal or glass and magnifies the play of colour as well as creating a protective surface. Assembled triplets typically use slivers of opal approximately 0.5 mm (0.02 in) thick.

Opal naturally contains water in its structure and may dry out if stored in dry conditions or subjected to heat or direct sunlight. It can often 'craze' in a fine network of cracks, which is sadly not reversible. The changing conditions following removal from the ground, and heat during polishing may cause this to happen. Opals from different locations have different stabilities based on their composition and water content. Opal dealers often hold back stock before selling to ensure they do not start to craze.

Wearing opal so it has contact with skin oils can help maintain its condition, and it should be stored in a cool environment with some humidity. Opal is safe to clean with a soft cloth and warm soapy water but is not suitable for ultrasonic or steam cleaners. It is porous and susceptible to chemical attack, so avoid contact with hairspray, makeup and perfume.

Occasionally opal is treated to enhance its durability and lustre. Due to its porous nature, it can be impregnated with oil, wax or polymer/plastic. This fills cracks, improves clarity and creates a harder surface. It may also be thinly coated with a polymer. A good way to test this is using a droplet of water to see if it is absorbed or sits on the surface. Opal may also be dyed to create a darker body colour imitating black opal. Opal composites are known, created using small pieces of opal set into black resin.

Opal is imitated by glass and plastic. However, the appearance, especially under magnification, is quite different. Opalite, known from the 1960s, is the trade name for a glass imitation that shows opalescence. It is transparent to milky with a bluish sheen caused by the scattering of light, typically inclusion free, or with gas bubbles. Slocum stone is a glass imitation included with

◀ Opal doublets of a precious opal slice on a black backing (left) and on a common opal backing (right).

◀ Slocum stone, a glass imitation of opal containing tiny iridescent sheets.

▼ A slab of artificially created Gilson opal, with the columnar structure visible on the side.

tiny iridescent sheets. These reflect light giving an attractive appearance similar to play of colour, but are easily distinguished by hand lens due to their crinkled look.

Synthetic opal was created by Pierre Gilson in 1974. Made from silica spheres, it is a close but not an exact counterpart of opal's chemical composition, structure or properties, thus is not a true synthetic. Gilson opal shows a play of colour but is brighter with distinct colour regions. Under magnification it has a regular polygonal pattern described as 'lizard skin'. Viewing side on reveals a columnar structure beneath the polygons, which is not seen in natural opal.

Peridot

Composition $(Mg, Fe)_2(SiO_4)$ magnesium iron silicate **Crystal system** orthorhombic **Hardness** 6.5–7 **Cleavage** poor **Fracture** conchoidal **Lustre** oily, vitreous **SG** 3.21–3.48 **RI** 1.634–1.710 **Birefringence** 0.032–0.038 **Optical nature** biaxial + **Dispersion** 0.020

Peridot is the transparent gem variety of olivine, enjoyed for its fresh citrusy yellow green to olive green colours. It is also a favourite of historians, as one of the oldest known gemstones, mined for thousands of years. This fascinating gem normally forms deep within Earth but is also one of the few to come from outer space, found within meteorites.

The history of peridot is somewhat ambiguous. First discovered and mined on Zabargad in the Red Sea (also known as St John's Island and Topazios Island) in at least 300 BC, and possibly as far back as 1500 BC, it was used by ancient Greeks and Egyptians who called it 'the gem of the sun' and was known by ancient Romans as the 'emerald of the evening'. Many believe Cleopatra's favourite gem emerald was actually peridot.

The naming of peridot has a similarly complex history. The Roman philosopher Pliny the Elder's *Historia Naturalis* refers to 'topazos/topazios', which were jewels of green to yellow, translated as topaz but now generally accepted to be peridot. 'Chrysos' meant golden, and 'chrysolithos/chrysolithias' was another term used for yellow gems. Some green gems named smaragdus (often interpreted as emerald) may have been peridot. Peridot, topaz and chrysoberyl have all been called chrysolite, but this name is generally no longer accepted to avoid confusion. The name olivine was applied in 1790 for the olive colours. The name peridot has been used for nearly 1,000 years and is thought to be from the Arabic 'faridat' meaning gem, or Greek 'peridona' meaning give in abundance. Today it is the accepted name for gem-quality olivine.

Olivine is not a mineral species, but a solid solution series between two closely related silicate minerals – magnesium-rich forsterite (Mg_2SiO_4) and iron-rich fayalite (Fe_2SiO_4). Forsterite when pure is colourless, occasionally faceted as a rare gem. As iron substitutes for magnesium, blending the composition towards fayalite, it causes a yellowish green to brownish green colour. Peridot's composition is closest to forsterite, and it can be considered an iron-bearing variety of it. The finest green colour occurs around 12–15% iron, becoming yellower or browner with increasing iron. Impurities of chromium or nickel are thought to produce a brighter green colour.

◀ A gem quality crystal of olivine variety peridot from Zabargad in the Red Sea.

With orthorhombic symmetry, crystals tend to be stocky or tabular with rounded wedge-shaped terminations. Twinning is common. It may also be granular, often forming as rounded clasts within volcanic rock.

Most peridot is formed deep in Earth within the upper mantle, brought to the surface by volcanic activity or plate tectonics. It is a common rock-forming mineral found in mafic and ultramafic rocks such as basalts and peridotites. Gem-quality crystals are found in a few locations around the world, typically in ancient lava beds as phenocrysts (nodules of granular crystals) within basaltic volcanic rocks. On rare occasions crystals form in veins or cavities.

Zabargad (St John's Island) in the Red Sea off the coast of Egypt was historically an important source. The island seems to have been lost over time, rediscovered in the past few hundred years, but producing low quantities. This small and unusual place is formed of a sliver of mantle uplifted to the surface by plate tectonics. Peridot is found in veins in peridotite, forming large crystals up to 20 cm (8 in). Zabargad has produced some of the finest and largest crystals and faceted gemstones. The largest known is a 311.8 ct faceted gemstone held in the Smithsonian Institution.

Peridot Mesa in San Carlos Apache Reservation in Arizona, USA, is today one of the major sources of peridot. The basalt tops are mined for nodules of granular peridot. Faceted stones in large sizes are rare, averaging 3 ct or less. Peridot is also found in New Mexico and Hawaii where green sand beaches are formed of olivine grains. Myanmar has produced some magnificent large crystals and faceted gemstones of rich saturated green colour, rivalling those from Zabargad. Colourless gemstones of forsterite also occur here. Sapat in Pakistan has wonderful crystals often found with magnetite, and produces high-quality gemstones of fine colour, although commonly with a yellowish tinge. China is also a large producer, with minor sources including Sri Lanka (olive green), Norway (minty green), Vietnam, Tanzania, Australia, Ethiopia and Brazil.

Perhaps most exciting is this gem's extra-terrestrial origins. Peridot occurs in pallasite meteorites as rounded crystals within the metallic iron-nickel matrix. These meteorites are very attractive when sliced and polished, allowing light through. Rarely, some crystals are gem quality, without too many fractures, and large enough to facet. The resulting gemstones are typically small and rarely seen on the market. These meteorites formed just a few million years after the formation of the solar system, making them the oldest gems in the world.

▼ A polished slice of a pallasite meteorite revealing transparent peridot crystals in metallic matrix, from the Atacama Desert, Chile.

▼ An exceptional 146.17 ct peridot from Zabargad.

The best quality peridot is eye clean with an oily look. Characteristic inclusions are 'lily pads' of disc-shaped halos (a curved stress crack) around a minute black crystal, typically octahedral chromite. Brown mica flakes can give stones a brownish look. Needle-like ludwigite and vonsenite inclusions are common in Pakistani peridot. Other inclusions are clear crystals and short, fine needles. Eye-visible inclusions, particularly dark crystal inclusions, lower the value. Parallel silky inclusions may give a chatoyant effect when cut en cabochon, and very rarely produce a four-rayed star.

Most peridot is yellowish green, with rich pure green the rarest and most desired. Larger gemstones tend to have a finer, more intense colour, also appearing more oily or greasy. With high birefringence up to 0.038, peridot shows a strong doubling effect, easily seen by eye in larger stones – a good identifying feature. Peridot is weakly pleochroic green/yellow-green.

Peridot is cut in a variety of shapes and styles, making best use of the colour and shape of the rough. Step, brilliant and mixed cuts are popular with round, oval and cushion shapes common. Gemstones tend to be under 5 ct, however more than 10 ct is not uncommon, and large transparent crystals occasionally yield gemstones more than 100 ct. Round beads, polished chips and cabochons are also fashioned.

This gem is reasonably durable with poor cleavage, however it can be brittle and its hardness of 6.5–7 is moderate, making it vulnerable to scratches and chips. Settings should be protective, particularly in rings. Peridot may be damaged by acids, and contact with hairspray, perfume and make-up should be avoided. It is safe to clean in warm soapy water but steam and ultrasonic cleaners are not recommended. Peridot is rarely treated but surface-reaching fractures may be filled to improve clarity. Older jewellery pieces can have foil backings on peridot to improve its colour.

Forsterite has been synthesized, grown by the crystal pulling method. It is coloured blue by adding cobalt, and is strongly pleochroic blue/violet/purple. It is used as an effective imitation of tanzanite, but can be detected by the different optical properties such as higher birefringence. Green synthetic forsterite or peridot is not seen on the market.

Peridot may be confused with other green gems such as tourmaline and demantoid garnet. It may be simulated by green synthetic corundum, spinel, quartz and cubic zirconia, or glass. Composite stones such as garnet-topped doublets of peridot colour are also known. All these can be distinguished by their differing physical and optical properties.

Quartz and its varieties

Quartz is one of the most common minerals in Earth's crust. As a gem, it has been used for adornment and carvings for thousands of years, celebrated for its diversity with more varieties than any other. Quartz is classified into two groups. Well-crystallized or macrocrystalline quartz creates large transparent gemstones, and includes varieties such as amethyst and citrine. Compact or cryptocrystalline quartz comprises tiny interlocking crystals and is known as chalcedony. Its colourful and often banded varieties include agate and chrysoprase.

Quartz is mentioned by Greek naturalist and philosopher Theophrastus in his book *Peri Lithon* (*c.* 314/315 BC) as 'kristallos' meaning icy, and it was common belief these clear crystals, now known as rock crystal, were eternal ice. The name quartz is thought to be derived from the term 'quertze', used by Saxon miners in the 1500s to describe a white vein stone. By the 1700s the name was established and the word crystal, from 'kristallos', applied to the angular form of any mineral.

Macrocrystalline varieties include pure colourless rock crystal, purple amethyst, golden citrine, smoky quartz and pink rose quartz. The colours are caused by chemical impurities or included minerals, and some colours are created by treatments such as heating and/or irradiation. These often-transparent varieties are faceted or carved, and rock crystal is popular for the range of inclusions encapsulated in its icy depths.

Cryptocrystalline quartz is a dense aggregate of microscopic interlocking crystals. Many varieties are grouped under the term chalcedony, and these are often considered to be separate from macrocrystalline quartz. They have been known and named since ancient times, and were only scientifically recognized as the same mineral as quartz in the late 1800s. Chalcedony occurs in botryoidal, stalactitic, nodular and banded forms, often lining geodes or cavities. Impurities create different colours, and varieties are primarily defined by their appearance, for example agate is banded, carnelian is orange and chrysoprase is green. They are popular for ornamental items and carvings, as well as cabochons and beads.

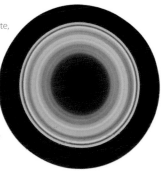

◀ Glassy crystals from the Alps are very popular with collectors, like this one from La Gardette, France.

▶ A polished agate slice with perfectly circular banding.

As the varieties of macrocrystalline and cryptocrystalline quartz are numerous, only the most well known in the gem trade are discussed here.

Quartz forms in almost all geological settings, including igneous rocks, particularly in granite and pegmatites, and many metamorphic rocks. As a stable and durable mineral, quartz weathers and accumulates as a major component of many sedimentary rocks such as sandstone. It also occurs as milky quartz veins, formed in fissures from silica-rich fluids, and is associated with important ore minerals such as gold.

Crystallizing with trigonal crystal symmetry, quartz typically occurs as well-formed, elongate six-sided prisms with characteristically uneven pyramidal terminations. It may also be squat, needle-like or tapered, and is usually striated perpendicular to the length. Quartz ranges from tiny crystals lining cavities to huge pieces weighing hundreds of kilograms. There are many interesting growth forms, and quartz is often twinned.

Quartz has a glassy vitreous lustre. It is durable and is the mineral representing 7 on Mohs scale of hardness. It has poor cleavage but can be brittle, with a conchoidal fracture similar to glass. The physical properties of

◄ Smoky quartz crystals from Switzerland with six-sided prisms and uneven pyramidal terminations.

cryptocrystalline quartz, or chalcedony, differ slightly due to its compact nature, and while the individual microcrystals have the properties of macrocrystalline quartz, the aggregate does not. The lustre is not always as bright and hardness is slightly lower at 6.5–7, but chalcedony is tougher and less brittle. The specific gravity varies depending on the porosity and the presence of water and impurities. As quartz is so common and durable, it not only forms a large proportion of sand but is also found in dust. Note that gem materials with a hardness of less than 7 may be scratched when wiping off dust.

With suitable hardness, toughness and high chemical resistance, quartz makes an excellent gemstone for use in jewellery. Clean in warm soapy water, and ultrasonic cleaners are safe to use unless inclusions or fractures are present. Steam cleaners and heat should be avoided as they may cause stones to fracture. Extra care is required for all coloured macrocrystalline varieties. As colour-altering treatments occur at relatively low temperatures, avoid heat such as a jeweller's torch as the colour may change or fade. Cryptocrystalline varieties may be porous, so avoid contact with household chemicals, perfumes and hairspray, especially if dyed.

Quartz is chemically pure and used extensively in industry. Sand is one of its main uses, as is gravel for concrete. It is mined for the manufacture of glass, abrasives and ceramics, and the silicon is used in silicon chips. Transparent rock crystal is used in electronics as an oscillator due to its property of piezoelectricity, whereby applying mechanical pressure generates an electrical charge. The reverse is also true, and applying an electrical charge makes quartz resonate at such a precise frequency it is used in battery-operated watches to keep time. It vibrates exactly 32,768 times per second and is more accurate than a pendulum clock. It is also used in radios and mobile phones.

Synthetic quartz has been grown commercially by hydrothermal methods since the 1950s, and used for many industrial processes including modern electronics and watches. It is also fashioned into gemstones in a range of colours. It may be detected by a lack of natural inclusions, instead containing breadcrumb-like inclusions, however most synthetic gems are eye clean and can be difficult to distinguish without laboratory analysis.

▶ Synthetic quartz crystal grown by a hydrothermal method.

Quartz, macrocrystalline

Composition SiO_2 silica **Crystal system** trigonal **Hardness** 7 **Cleavage** poor **Fracture** conchoidal **Lustre** vitreous **SG** 2.65 **RI** 1.544–1.553 **Birefringence** 0.009 **Optical nature** uniaxial + **Dispersion** 0.013

Rock crystal

Colourless rock crystal is the most common variety of macrocrystalline quartz. It is found and mined all over the world, occurring anywhere that quartz may form, with beautiful crystals extracted from pegmatites, cavities and hydrothermal veins. Brazil is the most important supplier, predominantly from pegmatite deposits in Minas Gerais. Other sources include Madagascar, Russia, the USA, India, Peru, Myanmar, Australia and the Alps.

Rock crystal can occur in huge transparent crystals allowing for flawless gemstones of hundreds and even thousands of carats. Faceted gemstones are bright, but lack fire due to low dispersion, so are used in costume jewellery or as affordable alternatives to diamond. Rock crystal is commonly carved or polished into spheres and other shapes. Individual crystals often have their faces polished, with small crystals set into jewellery such as pendants, and large ones sold as ornaments.

Thin slices of rock crystal are used in composite doublets or triplets to provide a durable cap on softer gems such as opal.

An incredible range of attractive inclusions makes rock crystal popular. As quartz is one of the last minerals to crystallize in geological environments, it encloses other minerals, as well as liquids and gas bubbles. Its physical durability and chemical resistance protects the encapsulated minerals, including fragile needle-like crystals, allowing them to survive geological changes over millions of years. The inclusions not only create appealing gemstones, but offer a window into the past.

▶ Glassy Alpine rock crystal from Austria.

Common varieties of included rock crystal are:

- **Rutilated quartz** – containing metallic needles or hairs of golden, silver, red or black rutile crystals. The needles may be in clusters, randomly crisscrossing, curved or in a six-rayed star around a silvery hematite centre. Rutilated quartz with a net-like, reticulated appearance, is known as sagenite. It is predominantly sourced from Brazil and India.

- **Tourmalinated quartz** – containing slender needles of black schorl tourmaline. The quartz ranges from transparent to milky, with randomly orientated schorl. Elongate crystals of different coloured tourmaline are more rarely seen.

- **Strawberry quartz** – with numerous elongate red inclusions, thought for many years to be lepidocrocite, now determined to be the iron oxide hematite.

- **Pink fire quartz** – containing tiny flakes of the copper sulphide mineral covellite. The flakes give an incredible metallic purple to pink iridescent flash as light reflects off them.

- **Dumortierite quartz** – containing inclusions of the borosilicate mineral dumortierite, creating an opaque to translucent, blue to violet gemstone. This is quite rare and highly valued.

Other mineral inclusions may be slender green actinolite, mossy green to brown chlorite, golden pyrite and other sulphide minerals, fluorite of various colours, and the rare copper silicates shattuckite and gilaite in tiny green to blue spherical clusters. Included rock crystal is fashioned to highlight the inclusions, into cabochons, faceted gemstones, and beads, or carved into ornamental objects. Milky white vein quartz containing flecks of native gold is sometimes polished for inlays or set into jewellery.

A metallic coating may be applied to rock crystal using thin film deposition. Titanium oxide creates a colourful iridescent effect sold as 'mystic' quartz, and a light blue sold as 'aqua aura' is created

▶ Golden rutile in a quartz cabochon, set in a ring.

▼ A faceted rock crystal with a single central schorl inclusion, perpendicular to the table, reflected internally by the pavilion facets to create a pinwheel effect.

◀ A quartz crystal coated with a molecular layer of gold, creating blue iridescence, known as 'aqua aura' quartz.

with gold. This treatment can be detected by
the unnatural colours and scratches in the softer
coating. Quartz may be heated and quickly
cooled, inducing many fractures which allow dye
to penetrate, known as quench crackled quartz.
This technique has been used to imitate other
gems such as emerald and is detected by colour
concentrations along the cracks. Rock crystal may
be irradiated to create smoky quartz, as it naturally
contains the aluminium impurities responsible for
the brownish-grey colour. Synthetic rock crystal is
not normally used for gemstones.

Rock crystal has been used to imitate diamond
for centuries, but is easily distinguished by its lack
of fire. It may be confused with colourless topaz,
sapphire and the lesser known but similarly water
clear phenakite, but distinguished by a lower
refractive index.

Smoky quartz

This brown to grey quartz variety is named for
its smoky appearance. It varies from clear to
translucent, and from very pale to almost black,
known as morion. The colour is caused by the
presence of aluminium impurities substituting
for silicon in the crystal structure, combined with
irradiation. This may occur naturally underground or
be artificially induced by gamma irradiation of rock
crystal. This treatment is not detectable, and very
dark smoky quartz should be treated with suspicion.

Smoky quartz occurs in granite, pegmatite and
metamorphic gneiss, which contain traces of
radioactive minerals. Crystals may be very large,
allowing faceted gemstones of thousands of carats.
Brazil is a leading source, known for large crystals, and
other sources include Madagascar, Mozambique,
Russia, Ukraine, Switzerland, Bolivia, Sri Lanka,
Myanmar, Australia, the USA and Spain. Smoky to
yellowish crystals from the Cairngorm Mountains
in Scotland are known as cairngorm. The natural
process that creates the smoky colour requires low
temperatures and a geologically long time, hence

▼ A 129.72 ct clear,
step cut, smoky quartz
gemstone from India.

smoky quartz is found at high elevations including
the Alps, Cairngorms and Rocky Mountains.

Smoky quartz is pleochroic yellow-brown/red-brown.
Most gemstones are eye clean, but when present,
inclusions may be fluid or two-phase, negative
crystals, rehealed fissures and rutile needles.

Transparent smoky quartz crystals are faceted into
gemstones or beads, and lesser quality material
is fashioned as cabochons. As gemstones can
be large, they make excellent centre stones for
pendants and earrings. Smoky quartz is also
carved or shaped into decorative objects such
as large spheres. Some gems may fade under
ultraviolet light and extended exposure to sunlight
should be avoided. Certain smoky quartz can be
heat treated to produce a yellowish-green citrine.
Smoky quartz may be confused with topaz but is
distinguished by a lower refractive index.

Amethyst

The beautiful violets of amethyst make it the most
desirable variety of quartz. Its name is derived
from the Greek 'amethystos' meaning not drunken,
and amethyst amulets have been used to protect
against the effects of wine since ancient times. It
has been treasured as a gemstone for thousands
of years, found as beads, carvings and set into
the jewellery of Ancient Egyptians, Greeks and
Romans. It was a symbol of royalty in the Middle

Ages and Renaissance times, found in many royal jewels and religious items. The British Crown Jewels feature large, faceted amethyst gemstones on the Sovereign's Orb, and on the Sceptre, positioned above the Cullinan I diamond.

Amethyst is found in many geological environments, occurring in well-formed crystals, but not to the same size or range of forms as rock crystal and smoky quartz. Amethyst frequently forms aggregates of short, terminated crystals called druse, well known as the crystals lining geodes. These rounded voids formed from gas bubbles trapped in lava flows around which the volcanic rock solidified. Silica-bearing fluids later move through the voids, allowing amethyst crystals to grow, lining the cavity. Geodes can occur in enormous sizes with elongate shapes of several metres. Amethyst can also occur in veins with milky quartz, with a layered chevron pattern sometimes called amethyst quartz.

Egypt is an ancient source of amethyst with indications of mining from at least 2000 BC.

▼ Amethyst druse on a thin agate layer lining a geode.

Germany, France and the Ural Mountains in Russia are all historically important sources, producing fine gemstones over the past 500 years. The discovery of huge amethyst deposits in first Brazil, then Uruguay, in the early 1800s, created a steady supply for the world market, but ultimately led to a significant drop in its value. Rio Grande do Sul (Brazil) and neighbouring Artigas (Uruguay) host geode deposits in volcanic rocks formed from vast ancient lava flows, known as flood basalts. These are the most commercially significant amethyst deposits in the world, also exploited for agate. Other sources of fine amethyst include Zambia, Madagascar, Namibia, Kenya, Bolivia, Canada, the USA, Mexico, Sri Lanka, Myanmar, India, Australia and South Korea. Crystal specimens, especially from Mexico and Namibia, are popular with mineral collectors.

Amethyst ranges from deep reddish-purple to bluish-violet to pale mauve. The colour is caused by impurities of iron substituting for silicon within the crystal structure, together with gamma irradiation of the crystal. Amethyst is weakly pleochroic, exhibiting bluish-violet/reddish-violet colours. Gems from Russia may exhibit an alexandrite effect, appearing purple in daylight and deep purplish-red under incandescent light. The colour of some amethyst fades with exposure to ultraviolet light, and prolonged exposure to sunlight should be avoided. Similarly, avoid heat as it may cause the colour to fade or alter.

Angular colour zoning related to the crystal shape is common, with colour often concentrated in the tips. Rough is cut to produce a uniform colour when the gem is viewed face up. Gemstones with intense colour and no visible zoning are the most valued. Amethyst known as Russian or Siberian is prized for its intense violet colour with red flashes. These names are often applied based on the colour regardless of source.

Faceted amethyst is normally eye clean. When present, the most common inclusions are known as tiger- or zebra-stripes, formed of numerous parallel zigzagging lines seen under magnification and considered to be diagnostic of amethyst. Their presence is not fully understood, potentially caused by fluid inclusions, healed fractures or twinning. Fingerprints/veils of fluid inclusions, and two-phase inclusions with gas bubbles, are also common. Iron oxides, as plates or needles, are the most prevalent mineral inclusions. Red elongated hematite, known as beetle legs for its appearance, is seen in amethyst (and strawberry quartz), and golden fibrous tufts of goethite form attractive inclusions. Most amethyst has repeated internal twinning of alternating thin lamellae. While not normally seen in a cut gemstone, this creates a pattern under the polariscope that can indicate natural versus synthetic origin and can also be used to identify citrine formed from heat-treated amethyst.

Amethyst is faceted into gemstones in a range of sizes and cutting styles including freeform and fantasy cuts. Gemstones may be large with exceptional stones of several hundred carats. Lesser-quality material is fashioned into cabochons, and polished or tumbled into beads. Amethyst is also used for carvings and decorative objects. Small natural crystals and sections cut from geodes are set in jewellery to display the pointed crystals.

Several treatments are used on amethyst. A metal coating may be applied using thin film deposition to produce iridescence, known as rainbow amethyst. It may be dyed to produce a deeper colour and, rarely, is fracture-filled to improve the clarity. Amethyst is commonly heat treated to improve or change the colour. Gentle heating at low temperatures lightens deep colours, but cannot be used to darken pale ones. Heating to over 350–450°C (660–840°F) bleaches the purple colour to colourless or pale yellow. Increasing temperatures to around 560°C (1,040°F) then deepens the colour to reddish browns,

▲ Richly coloured 33.86 ct faceted amethyst from Uruguay.

▼ Tiger-stripe inclusions in amethyst. FOV 3.12 mm.

becoming milky white over 600°C (1,110°F). The yellow to brown gems are normally sold as citrine, sometimes known as burnt amethyst. Amethyst from certain localities will remain colourless at 400–500°C (750–930°F) or become light green prasiolite. Heat treatment is normally stable and can be undetectable. Irradiation of most heat-treated amethyst will return the original colour, but may also induce smoky colours.

Amethyst has been synthesized by the hydrothermal method since the early 1960s, commercially available since the 1970s, and synthetic amethyst is common on the market. It lacks natural inclusions such as tiger-stripes, colour zoning, and repeated twinning, however when eye clean it can be difficult to distinguish without laboratory analysis.

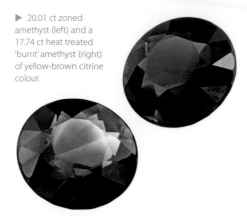

▶ 20.01 ct zoned amethyst (left) and a 17.74 ct heat treated 'burnt' amethyst (right) of yellow-brown citrine colour.

It is possible to confuse amethyst with purple sapphire, which was formerly known as oriental amethyst and differentiated only in the mid 1700s. The term amethystine is sometimes applied to purple gem materials such as glass. Amethyst may be confused with purple spinel and purple garnet, such as rhodolite, although both can be distinguished by their isotropic nature. Paler material may be confused with spodumene variety kunzite and pink scapolite. Tanzanite, and synthetic corundum doped with vanadium to produce a purplish-pink to slate blue colour change used to imitate alexandrite, can both appear similar to the pleochroic colours of amethyst. All can be distinguished by higher refractive indices.

Citrine

Citrine is the golden variety of quartz, ranging from pale yellow to brownish-orange. It is a popular gemstone for its warm colour, durability and affordable price. The name is derived from the French 'citron' and Latin 'citrus' for its lemon colour. Naturally coloured citrine is rare, and most gems on the market result from the heat treatment of amethyst, and occasionally smoky quartz. Citrine has been used since ancient times for jewellery, and some historical engraved gems, known as intaglios, have indications of heat treatment, suggesting that the process of heat treating amethyst to obtain citrine may have been used for more than 2,000 years.

This variety occurs in pegmatite and hydrothermal veins. Natural citrine, while rare, is found in many countries including Brazil, Bolivia, Spain, Madagascar, the USA, Zambia, Russia, Norway and Australia. Sources of amethyst for heat treatment include Brazil and Uruguay. All citrine geodes are heated amethyst, and care should be taken as the treatment makes them more fragile than amethyst geodes.

The colours of citrine are wide ranging. Natural citrine is pale yellow, greenish-yellow, yellow-orange through to brownish colours where it grades into smoky quartz. Citrine produced by heat treatment

◀ The colour range of citrine gemstones.

◀ A fantasy cut 20.45 ct lemon citrine with concave facets on the rear, likely irradiated then heat treated to produce the colour.

▶ A 133.32 ct faceted citrine with pale colour banding from Salamanca, Spain.

has a richer colour and a reddish overtone not seen in natural stones. Natural citrine is weakly dichroic yellow/brighter yellow, while heat-treated citrine lacks pleochroism. Different colours are given names – orange to reddish-browns are called Madeira citrine, golden colours are golden citrine, and heat treatment of amethyst from the Palmeira Mine in Brazil produces a fine orange colour known as Palmeira citrine. The most desirable colours are deep reddish-orange or intense yellow without brownish overtones, and lemon. Citrine's colour has several causes, which are not yet fully understood. These are: aluminium substituting for silicon combined with irradiation (similar to smoky quartz); iron substituting for silicon in the crystal structure; and iron combined with heat leading to the formation of submicroscopic iron oxide particles. As amethyst naturally contains iron impurities, heat treatment can create citrine. The colour change can be reversed by irradiation. Heat-treated amethyst is sold as citrine, and while the treatment should be disclosed, this does not always occur.

Smoky quartz (either naturally occurring or artificially irradiated colourless quartz) that also contains lithium can produce greenish-yellow colours when gently heated below 280°C (535°F), traded as lemon quartz, lime citrine or ouro verde (Portuguese for green gold). It is pleochroic

yellow/yellow-green and does not have the reddish overtones seen in heated amethyst. This material is mostly sourced from Brazil.

Most citrine gemstones are eye clean. Citrine formed by heat treating amethyst contains the same inclusions as amethyst – two-phase inclusions, other minerals, diagnostic tiger-stripes and repeated twinning. Citrine can form large crystals, and huge flawless gemstones of thousands of carats are known. It is faceted in a range of sizes and cuts, including particularly popular fantasy cuts. As gemstones are commonly up to 20 ct, citrine is used as an attractive centre stone in pendant, earring and ring settings. Lesser material may be polished or tumbled into beads. Citrine is also carved.

Synthetic citrine is grown by hydrothermal methods, with impurities of iron to create intense colour. It lacks the natural inclusions and colour zoning of natural citrine but may have breadcrumb inclusions. Citrine is often confused with golden-yellow topaz and reddish-orange imperial topaz. As topaz is a more valuable gemstone, citrine (particularly heated material) is often marketed as topaz quartz, gold topaz, and brownish-red gems as Madeira topaz. These names are misleading and should not be used. Other similar gems include yellow sapphire, golden beryl variety heliodor, yellow scapolite and yellow apatite. Citrine of deeper orange to red colours can resemble some garnets. They can be distinguished by different refractive indices and other optical properties.

Ametrine

This bi-colour variety of quartz has distinct purple and yellow zones, named for the combination of amethyst with citrine. Naturally coloured ametrine has one significant source – the Anahi Mine, Santa Cruz, Bolivia – and has been available on the market since the 1970s. The crystals occur in hydrothermal veins in fractured dolomitic limestones. When an ametrine crystal is cut perpendicular to the crystal length it may show six alternating sectors of purple and yellow, which resemble a radiation warning sign. What mechanisms or geological conditions cause this sector colour zoning is still uncertain, although it is suggested the citrine sectors have a higher iron and/or water content. Natural material may be heated and irradiated to produce the two colours, and ametrine may be grown synthetically. Consequently, this gem is not particularly rare on the market. Gemstones are transparent and common up to 30 ct. Ametrine is carved, faceted into bicolour gemstones such as elongate emerald cuts, or fancy cut with artistic use of the colour. Synthetic ametrine has been produced since 1994 and can be difficult to distinguish from the natural gem. The amethyst sectors normally lack the repeated twinning lamellae of natural amethyst.

▲ A 10.50 ct ametrine from Bolivia, fancy cut to make best use of the colour zoning. Cut by John Dyer Gems.

Prasiolite

Prasiolite (not to be confused with prase, prasolite or prase opal) is the transparent green variety of macrocrystalline quartz. It is named for its colour, derived from the Greek 'praso' meaning leek. It very rarely occurs in nature, known from just a few locations including Canada, Namibia and Poland. Almost all gemstones on the market are heat treated amethyst from specific localities, including the Montezuma Mine in Minas Gerais, Brazil. Heating to 400–500°C (750–930°F) produces the light green colour, and this treatment is stable. These gemstones have a colour similar to peridot. Prasiolite can be further treated by gamma irradiation then heat to create blueberry quartz of a deep violet-blue, resembling tanzanite. A dark green quartz, different to prasiolite, is sometimes marketed under green amethyst. This is created by irradiating amethyst with a high water content, from the geodes in Rio Grande do Sul, Brazil. It can be distinguished under the Gem-A Chelsea Colour Filter appearing red, while prasiolite remains green.

▼ An 11.64 ct pale green prasiolite, created by heat treatment of amethyst, probably from Montezuma, Minas Gerais, Brazil.

Rose quartz

This is one of the rarer macrocrystalline quartz varieties, coloured soft pink to peach. Rose quartz occurs in very light to medium tones, with pure pinks or purplish pinks most desirable.

◀ Rare, well-formed crystals of rose quartz on quartz, from Taquaral, Minas Gerais, Brazil.

▶ A 25.55 ct rose quartz, with milkiness caused by microscopic inclusions, from Minas Gerais, Brazil.

Rose quartz is found in granitic pegmatites and hydrothermal veins, in many locations around the world. Brazil and Madagascar are the main sources, and it is also found in the USA, Russia, Sri Lanka, Myanmar, India, Afghanistan, Namibia, South Africa and Mozambique.

There are two types of rose quartz. The more abundant type occurs as translucent masses, rather than well-formed crystals. It has a uniform milky pink colour, is opaque to translucent and rarely transparent. The cause of the rose colour is much debated. Current research indicates that microscopic fibrous inclusions of a mineral related to dumortierite create the colour and hazy appearance. This rose quartz is dichroic with two shades of pink, caused by a preferred alignment of the fibres. The second type is much rarer, occurring as well-formed crystals in attractive clusters or rosettes on white quartz. It is commonly transparent without a milky nature, and may show colour zoning. The cause of this rose colour is impurities of aluminium and phosphorus within the crystal structure, combined with irradiation. As this quartz differs mineralogically it is sometimes called pink quartz. It is found in only a few locations, mainly the phosphate pegmatites near Galiléia, Minas Gerais, in Brazil. This pink quartz is very light and heat sensitive, and the colour fades quickly with exposure to ultraviolet light, although it can be returned through irradiation. Gemstones of this type should be stored in the dark and worn in the evenings to protect the colour.

The more abundant translucent to opaque rose quartz is carved, fashioned into larger decorative objects such as spheres, and is used for cabochons, beads and faceted gemstones. The rarer transparent material occurs in smaller crystals than other quartz varieties, therefore finished gemstones are also much smaller. Transparent gems are normally under 30 ct, with better colour seen in larger stones.

Although most rose quartz is milky due to microscopic fibrous inclusions, it does not normally contain other inclusions visible by eye. If the fibrous inclusions are aligned, and it cut en cabochon or in a sphere, rose quartz may show asterism with a six-rayed star or chatoyant cat's eye effect. Curiously, the star effect can be seen by reflected and transmitted light. Very rarely massive rose quartz

▼ A 157.39 ct rose quartz sphere showing asterism with transmitted (not reflected) light. From Maine, USA, 2.8 cm (1 in) diameter.

▼ Polished cabochon of Hawk's eye quartz from Orange River, South Africa, showing chatoyancy on the wavy banding.

displays an unusual optical phenomenon known as Tyndall scattering. Light entering the gemstone is scattered by minute inclusions of similar size to the wavelength of visible light. As blue wavelengths of light are scattered more than other colours, it can give the stone a blue appearance.

Synthetic rose quartz is grown hydrothermally, with the addition of aluminium and phosphorous, followed by irradiation to produce the pink colour.

Rose quartz can be confused with the pale pink colours of beryl variety morganite and spodumene variety kunzite, but may be distinguished by differing optical properties.

Chatoyant quartz – cat's eye, tiger's eye, hawk's eye

These translucent to opaque quartz varieties exhibit magnificent silky chatoyancy caused by their fibrous internal structure. Cat's eye quartz occurs variously in yellow, brown, green and rarely blue, containing numerous aligned fibres or needle-like inclusions. The inclusions may be asbestiform minerals such as crocidolite or green actinolite. When cut en cabochon a cat's eye effect is seen, although

▲ Tiger's eye cabochon set in a gold pin, exhibiting a cat's eye effect. From South Africa.

more diffuse than cat's eye chrysoberyl. Cat's eye quartz is sourced from Sri Lanka, India and Brazil.

Tiger's eye is well known for its golden and chocolate brown stripes, while the lesser known hawk's eye (or falcon's eye) offers pale bluish-greys. These quartz varieties fall between macro- and cryptocrystalline quartz. They contain the mineral crocidolite (blue asbestos), which has been silicified but retained its fibrous structure. Hawk's eye contains partially silicified crocidolite with dark blue to grey stripes. Tiger's eye is fully silicified, with the crocidolite replaced by quartz and iron oxides (limonite and/or goethite) creating golden to brown colours. Both form in iron ore deposits where crocidolite grows in thin veins, orientated perpendicular to the vein edges. Most tiger's eye and hawk's eye comes from South Africa, with minor amounts from Australia. Both take a good polish with vitreous lustre, used as cabochons, beads, tumbled stones, decorative items and flat slices for ornamental stone. The chatoyant effect is spectacular, moving across the fibrous waves, and the alternating colours appear to reverse as the stone is turned. Tiger's eye has a brighter chatoyancy than hawk's eye and may also be bleached to lighten the colours. While crocidolite is well known for being hazardous, hawk's eye and tiger's eye are safe gem materials as the fibres are embedded within quartz, and silicified.

Aventurine quartz

The metallic glittery effect of aventurine quartz is due to inclusions of tiny flakes, a phenomenon known as aventurescence. It is most commonly dark green, included by fuchsite, the chromium-rich variety of muscovite mica. Less commonly iron oxides such as hematite impart a reddish to golden brown sparkle. The name aventurine, and aventurescence, are derived from the Italian 'a ventura' meaning by chance, following the supposedly unexpected creation of a shimmery artificial glass with copper inclusions called

◀ Aventurine quartz containing iron oxides, polished into a 13.1 ct heart shaped cabochon, from Nevada, USA.

goldstone or aventurine glass. Aventurine is commonly classed as a macrocrystalline quartz, however its compact granular texture of interlocking quartz crystals technically makes it a rock. It is sometimes classified as cryptocrystalline quartz, however the crystals are not generally microscopic. Aventurine is sourced from Brazil, India, Russia, Tanzania, Austria and Norway. Rich green aventurine may be confused with emerald and jade, but can be distinguished under magnification by the included flakes. Brownish colours may be confused with aventurescent feldspars such as sunstone, and artificial goldstone.

Quartz, cryptocrystalline

Composition SiO_2 silica + impurities + water
Crystal system trigonal (microcrystalline)
Hardness 6.5–7 **Cleavage** none **Fracture** uneven, splintery **Lustre** vitreous, waxy **SG** 2.55–2.91 **RI** 1.530–1.553 **Birefringence** 0.004–0.009 **Optical nature** uniaxial +
Dispersion none

Chalcedony

The term chalcedony encompasses the many different varieties of the compact form of quartz. The most popular are listed below. Chalcedony occurs as layers within cavities and veins, in rounded botryoidal clusters and stalactitic forms. It forms from silica-rich gels, close to Earth's surface at low temperatures. The silica is sourced from the weathering of rocks, and chalcedony is common in volcanic and sedimentary rocks.

Scientifically, the term chalcedony is applied to cryptocrystalline varieties where the microscopic crystals grow parallel in a fibrous structure, which is orientated perpendicular to banding or layers. This definition excludes jasper, which is microgranular in texture. In the gem trade, the term chalcedony

is normally used for milky bluish to grey material that does not fall under any other variety. The blue colour is caused by Rayleigh scattering of light, a phenomenon where light entering is reflected and scattered from tiny included particles which are much smaller than the wavelengths of light. As the shorter blue wavelengths are scattered more than other colours, the chalcedony appears blue. This same phenomenon explains why the sky appears blue, as sunlight is scattered from gas molecules in the air. Rayleigh scattering is similar to Tyndall scattering but from smaller particles. Blue varieties are sourced from the USA, India, Brazil, Uruguay, Madagascar, Namibia and Russia.

▶ A prism shaped bead of chalcedony, with a brownish body colour but appearing blue due to the scattering of light.

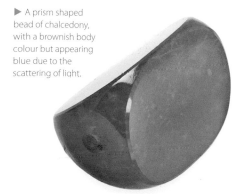

Chert, and its nodular form flint, are compact quartz-rich rocks composed of cryptocrystalline quartz (both as fibrous chalcedony and microgranular jasper) and impurities. They are historically and culturally significant, prized for their toughness and worked into sharp-edged tools. Arrow-heads are sometimes worn as pendants. Both take a good polish and are tumbled for beads. Chert is typically white to grey, sometimes red or green. Flint is generally darker in colour, and nodules are found in chalk or limestone formations. When struck against steel, flint produces sparks that can be used to light fires.

Agate

Agate is the translucent banded variety of chalcedony, with layers of varying thickness, colour and translucency. The name agate was first used by Greek philosopher Theophrastus in *Peri Lithon* (*c.* 314/315 BC), derived from the Achates river (now Drillo) in Sicily, along which it was said to have been discovered. Agate forms as linings

or fillings in cavities. Most occur as nodules in volcanic rocks, ranging from 1 cm (0.39 in) to more than 50 cm (20 in) in diameter, but some form in sedimentary rocks or hydrothermal veins. The exact process of agate formation is still debated. One theory is rhythmic crystallization from silica-rich gels. Agate grows inwards from the void's surface in concentric bands, horizontal layers, or a mix. It may fill the cavity or form a geode with an open centre, often with a final layer of sparkling quartz or amethyst druse. Horizontally layered agate is used in situ as a tiltmeter to measure post-formation movement of the host rock, by the tilt angle of the layers from horizontal.

Egypt is a historical source of agate, but it occurs all over the world. It has been mined and worked in the Idar-Oberstein region of Germany since at least 1548. The discovery of the vast agate and amethyst deposits in Brazil, in the early 1800s by German emigrants, led to a steady supply shipped directly to Idar-Oberstein. Today Idar-Oberstein is still the most important gem-cutting centre for agate. The most significant deposits are in Brazil and Uruguay. Other sources include Argentina, Mexico, Morocco, Madagascar, the USA, Australia, India, China and Scotland. Agate is widely used as a gem material including for cabochons, beads, carvings, ornamental objects, and pestle and mortars. Straight layers are perfect for engraving cameo and intaglio gems, utilizing contrasting colours.

▼ A polished slice of a prized Laguna agate nodule, from Chihuahua, Mexico.

Agate can form in any colour, with different coloured layers caused by different mineral impurities. Grey and white are the most common, and are more pure. Brown, red and yellow are caused by iron oxides. The layers have different translucency, which often appears wonderfully reversed when viewed in transmitted versus reflected light. Agate is porous and

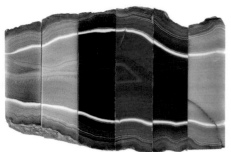

◀ Series of agate sections showing natural (leftmost) and artificial dyed colours.

▼ Delicately carved moss agate cup.

commonly dyed or stained in all colours, including unnaturally bright blue, pink and green. Different layers absorb dye differently, highlighting the banding. Dyes may not be stable, and should always be disclosed.

Types of agate are defined by their colour and pattern, inclusions, or source:

- **Fortification agate** – with angular banding that resembles the plan of a medieval fort.
- **Eye agate** – round concentric rings, often fashioned with a central point like an eye.
- **Lace agate** – scalloped or zigzag banding that resembles lace work. Blue lace agate occurs in Namibia, and crazy lace agate from Mexico has bright, often red, complex patterns.
- **Fire agate** – a shimmering rainbow iridescence on brownish botryoidal subsurface layers. The effect is due to fine layers of chalcedony with coatings of iron oxides such as goethite. Light is diffracted from the very thin layering, creating the iridescent rainbow effect. Cabochons are asymmetric, following the irregular patchy layers, in glowing reds and oranges to all rainbow colours. Fire agate is found in Mexico and Arizona, USA, and fine stones can rival opal.
- **Dendritic agate, mocha stone and moss agate** – with branch-like inclusions (dendrites) set in translucent colourless, pale yellow or grey backgrounds. Although lacking the banding of true agates, they are traditionally accepted as agate varieties. Dendritic agate has black,

brown or red dendrites of iron and manganese oxides, and is sourced from Brazil, India and the USA. Mocha stone is dendritic agate originally found in Mocha, Yemen. It has dark manganese-rich dendrites that resemble trees or landscapes in a yellowish background. Moss agate contains mossy green inclusions of chlorite or hornblende, which may oxidize to browns and reds. It is well known from India, and other sources include the USA, Canada, China and Scotland. Dendritic agate is carefully polished to bring the inclusions to the surface, and stones resembling trees or landscapes are highly valued. Moss agate is used in thin slices to highlight the inclusions, as cabochons or ornamental objects.

▼ A mouse carved from a single piece of agate utilising the banding (8.2 cm, 3.2 in high), next to a polished section of the same agate. Carved by Gerd Dreher.

Onyx

Onyx is a black or straight-banded chalcedony, so can be classed under agate, but is often considered its own variety. It is usually black or dark brown with white banding. Onyx is traditionally used for cameos with the image carved in white relief against the black background. The black layers are often opaque and used on their own. Agate is commonly dyed to create onyx. It is soaked in sugar solutions then heated in sulphuric acid, which deposits black carbon in the pore spaces to create black or dark layering. This treatment has been used since ancient times and the treated agate is difficult to distinguish from natural onyx. Onyx should not be confused with the unrelated marble onyx – a much softer, brownish, banded calcite marble used for decorative stone and carvings.

Sard

Sard is translucent brown or brownish-orange chalcedony, distinguished from carnelian by the browner hues. The colour is caused by impurities of iron oxides. Sard is not banded. It occurs as nodules within volcanic rocks such as basalt, and is found in the same environments as chalcedony and agate, sourced from India, Brazil and Uruguay.

Sardonyx

Sardonyx, as the name suggests, is a mix of sard and onyx with brownish-orange or red and white banding.

◀ A cabochon of onyx with straight black, brown and white banding.

Carnelian

Carnelian, or cornelian, is translucent orange to reddish-orange chalcedony, coloured by impurities of iron oxides, particularly hematite. Carnelian is not banded, although the colour distribution may be cloudy. Uniform bright red and orange colours are most valued. Carnelian occurs as nodules within volcanic rocks, typically basalt, in the same environments as chalcedony and agate. It is sourced from Brazil, Uruguay and Egypt, with fine material from India. It is distinguished from sard by the orangey-red colour. Carnelian with brownish or yellowish overtones is heat treated to create redder colours. Other chalcedonies, such as agate, are dyed and heat treated to create the carnelian colour, distinguished by their banding, revealed in transmitted light.

◀ Carnelian carved bowl from the Sir Hans Sloane collection, Natural History Museum, London dating from pre-1750s.

Chrysoprase

Chrysoprase is the minty to apple green variety of chalcedony, and the most valued. The name is derived from the Greek 'chryso' meaning golden and 'praso' meaning leek, alluding to its bluish- to yellowish-greens. Chrysoprase ranges from near transparent to opaque, most desired with translucent saturated uniform hues. It is coloured by microscopic inclusions of nickel-rich minerals, with traces of iron creating yellowish overtones.

It forms in cavities or veins within altered nickel-bearing mafic rocks such as serpentinites, or in overlying laterites. Historically sourced from Poland, most chrysoprase now comes from Marlborough in Queensland, Australia, known for its fine translucent green colour. Other sources include Tanzania, Brazil, India, Madagascar, Western Australia, Russia and the USA. The colour of some chrysoprase fades with strong light or heat, but is reported to be restored through rehydration. Chrysoprase is an effective imitation of jade for its colour, lustre and toughness. It has been imitated by chalcedony that is dyed or stained using chromium or nickel, and green glass, the latter distinguished by bubble and swirl inclusions. Prase opal is a common opal of the same colour, also coloured by nickel. It is found at some of the same locations but is distinguished by a differing refractive index and specific gravity. Lemon chrysoprase is a yellow-green gem material found in Western Australia, composed of quartz and magnesite, with or without chrysoprase.

Prase and other green cryptocrystalline quartz

There are many green cryptocrystalline quartz varieties, named for their appearance. Some, such as prase, plasma and bloodstone, have ambiguous definitions, classed as chalcedony in some references, and jasper in others. Prase is the name applied to medium to dark leek-green quartz, coloured by inclusions of minerals such as actinolite or chlorite. The name originates from the Greek 'praso' meaning leek. The use of the name has changed over time, originally describing a green quartzite rock, then cryptocrystalline green quartz including opaque green jasper and translucent green chalcedony. Today the name also encompasses macrocrystalline green quartz included by actinolite. Prase is different to prasiolite (light green transparent quartz), prasolite (a leek-green variety of chlorite), and prase opal (a bright green common opal coloured by nickel).

Plasma is a dark green cryptocrystalline quartz. The name is applied to opaque green jasper, and translucent green chalcedony, although this is normally darker and more opaque than prase. The colour may be caused by inclusions such as chlorite, and it sometimes has yellowish or whitish spots. Sources include India, Madagascar, Egypt and the USA. Bloodstone, also called heliotrope, is opaque dark green from microscopic inclusions of actinolite or chlorite, with red flecks resembling spots of

◀ 45.8 ct cabochon of translucent chrysoprase.

▶ Bloodstone carved into a cameo.

blood caused by iron oxides. It has been revered as a talisman due to the association with blood, vitality and bravery. It is found in many places including India, Brazil, the USA, Australia and China. Chrome chalcedony, also traded as mtorodite, is a green chalcedony from Zimbabwe coloured by chromium. It closely resembles chrysoprase, but may be distinguished under the Gem-A Chelsea Colour Filter by appearing red, while chrysoprase remains green. Chrome chalcedony was a favourite in Roman times, sourced from Turkey.

Jasper

Jasper is a microcrystalline variety of quartz defined by its opacity and granular, rather than fibrous, texture, so is scientifically excluded from the umbrella of chalcedony. It is sometimes considered a variety of chert. It has a high content of mineral impurities and is commonly coloured red, yellow and brown by iron oxides. It is frequently multi-coloured with a diverse range of distinctive patterns. The name is derived from the Old French 'jaspre' for speckled stone. It has been a popular gemstone for

▶ A polished cabochon of mookaite from Mooka Creek, Western Australia.

▼ Ocean jasper cabochon from Madagascar.

millennia, used for carving, engravings, ornamental stones and inlays, cabochons and beads.

This variety forms in layers in metamorphic and sedimentary rocks, such as banded iron ore formations. It occurs in veins and cracks in volcanic rocks, associated with chalcedony and agate but not as geodes. Occasionally it forms veins in igneous rocks such as granite. It is found around the world, including India, Australia, the USA, Egypt, Brazil, Uruguay, Canada, Mexico, Madagascar and Russia.

There is a huge range of jasper varieties, named for their appearance or the location in which they are found:

- **Orbicular jasper** – with concentric rings, some with the appearance of round 'eyes'. Leopard jasper has tan rings, and ocean jasper has multi-coloured, often dark green, rings hailing from a coastal location in Madagascar.
- **Egyptian jasper** – a deep red to yellow jasper, and a light brown with fine black bands.
- **Moss jasper** – the opaque version of moss agate with inclusions of hornblende.
- **Picture jasper** – appears to depict scenes or landscapes, with lovely specimens from Biggs Junction, in Oregon, USA.
- **Agate jasper** – opaque banded jasper.
- **Mookaite** – comes in warm reds, yellows, browns and mauve from Mooka Creek in Western Australia.

Jasper may be dyed, which is stable and detectable, or heat treated to change the colour which is stable but not detectable.

Petrified wood

Petrified, or fossilized, wood is where the wood cells have been replaced by silica, but retain their cellular structure. This may occur by any of the cryptocrystalline quartz varieties including agate and jasper. Fossilized bone and corals are also known to be replaced by chalcedony.

Rhodochrosite

Composition $MnCO_3$ manganese carbonate
Crystal system trigonal **Hardness** 3.5–4
Cleavage perfect **Fracture** conchoidal, uneven
Lustre vitreous, pearly **SG** 3.40–3.72
RI 1.578–1.820 **Birefringence** 0.201–0.220
Optical nature uniaxial - **Dispersion** 0.015

▲ Attractively banded polished slab of 'Inca Rose' rhodochrosite from Capillitas, Argentina.

▼ Rhodochrosite crystal specimen 7 cm (2.75 in) wide from the Kalahari manganese fields, South Africa.

Rhodochrosite is recognized by its attractive raspberry red and pale milky banding. It is not common in the gem trade, and even rarer are the magnificent, transparent, intense pink faceted gemstones. The name is derived from the Greek 'rhódon' meaning rose and 'chro' meaning coloured, and the colour is derived from its manganese.

This mineral is normally opaque to translucent in fine-grained massive to banded aggregates. It occurs as vein, stalactitic or nodular formations exhibiting curved irregular banding. Rhodochrosite is part of the trigonal crystal system and on rare occasions forms transparent crystals, with rhombohedral or scalenohedral forms dominant. Crystal specimens and faceted transparent gemstones are highly valued by collectors.

It forms in low to moderate temperature hydrothermal veins, and is also found in sedimentary manganese deposits or metasomatic deposits. The Capillitas mining area in Argentina has been the main source of banded and stalactitic material for many decades. It is the national gemstone, also known as Inca Rose. Well-formed crystals occur from Sweet Home Mine in Colorado, USA, where it is the state gemstone. Other sources include the Kalahari manganese fields in South Africa, Wutong Mine in Guangxi, China, and several locations in Peru, as well as in Germany, Romania, Mexico, Japan and Brazil.

The banding occurs in beautiful dark to light pink, white or yellowish layers, with serrated or lace-like borders. Material fashioned from stalactitic growths exhibits beautiful concentric

banding when sliced. Rhodochrosite is used as an ornamental stone, or fashioned into cabochons and beads to highlight the distinctive markings. Rare faceted gemstones range from deep red, rose pink to orangey pink. They may be eye clean, or with two- or three-phase inclusions and partially healed fractures. A strong doubling effect can be seen due to the high birefringence. Rhodochrosite is a fragile gem material with a low hardness of 3.5-4, and perfect rhombohedral cleavage in three directions. This, combined with its relative rarity, makes it a collector's gemstone. It has a vitreous lustre, which may be pearly on

▲ A 3.64 ct step cut rhodochrosite from Sweet Home Mine, Colorado, USA.

cleavage planes. Opaque material will show variations in the lustre and quality of polish across the banding.

Rhodochrosite should be treated with care. It is best worn in protected settings such as pendants and earrings, and should be stored separately to prevent scratches. As a carbonate mineral it reacts with acids, so avoid contact with household cleaners, perfume and hairspray. It is also slightly soluble in water, so wipe clean with a damp cloth. Rhodochrosite has a similar appearance to rhodonite but may be distinguished by lower hardness, rhombohedral cleavage and banding with serrated edges.

Rhodonite

Composition $CaMn_3Mn(Si_5O_{15})$ manganese silicate **Crystal system** triclinic **Hardness** 5.5–6.5 **Cleavage** perfect **Fracture** uneven, conchoidal **Lustre** vitreous, pearly **SG** 3.40–3.76 **RI** 1.711–1.752 **Birefringence** 0.010-0.014 **Optical nature** biaxial + **Dispersion** none

Rhodonite is a manganese silicate mineral that derives its appealing rose-pink colour from manganese. Its name comes from the Greek 'rhódon' meaning rose. It forms in the triclinic crystal system and occurs primarily as opaque to translucent compact or granular masses of deep to pale pink. Black dendritic inclusions of manganese oxides are a common and distinguishing feature, giving an attractive appearance. When well crystallized, rhodonite occurs in tabular or elongate prismatic crystals. Very rare transparent gem-quality material is faceted into superb rose pink, red or brownish red gemstones, with distinct red to orange pleochroism.

Rhodonite is found in manganese and calcium-rich skarns, a metasomatic rock that forms at the contact between a silicate rock or magmatic melt, and a carbonate rock. Massive material is mined in the Ural Mountains in Russia and in Tanzania. Gem-quality crystals are known from Broken Hill In Australia, Morro da Mina Mine in Brazil, also China and Peru. Other locations include New Jersey and California in the USA, Japan, Madagascar, Mexico and Sweden.

This gem is moderately hard, registering 5.5–6.5 on the Mohs scale. It has perfect cleavage in two directions at almost 90 degrees, and requires some

▼ Gemmy red crystals of rhodonite with silvery galena from Broken Hill, NSW, Australia.

▼ A decorative box made of rhodonite with typical black veining from Cavnic, Romania.

care to avoid scratches and knocks. Lustre is vitreous to dull, or pearly on cleavage planes. Opaque material, popular for the contrasting pink and black, is fashioned into cabochons and beads, and used as a decorative stone for ornaments, carvings and tiles. Beautiful transparent dark red crystal specimens and faceted gemstones are sought by collectors, but rarely seen in jewellery due to their fragile nature. Rhodonite may be distinguished from the similar looking rhodochrosite by its higher hardness and black veining.

Scapolite

Composition $Na_4Al_3Si_9O_{24}Cl$ to $Ca_4Al_6Si_6O_{24}(CO_3)$ sodium calcium aluminium silicate **Crystal system** tetragonal **Hardness** 5–6 **Cleavage** good **Fracture** conchoidal **Lustre** vitreous **SG** 2.50–2.78 **RI** 1.531–1.600 **Birefringence** 0.004–0.037 **Optical nature** uniaxial - **Dispersion** 0.017

▼ A 4.5 cm (1.8 in) high scapolite crystal.

Scapolite is known for attractive yellow, pink and violet gemstones, and fine cat's eyes, but despite the beautiful colours tends to be a collector's gemstone. Mineralogically, scapolite is a mineral group forming a solid solution series between sodium-rich end member marialite ($Na_4Al_3Si_9O_{24}Cl$) and calcium-rich end member meionite ($Ca_4Al_6Si_6O_{24}CO_3$). The term wernerite is used for intermediate members with a blend of the two compositions, or strongly fluorescent scapolite. As it is difficult to visually distinguish the two end members, and composition is typically mixed, gemstones are normally called scapolite. The name is thought to come from the Greek for shaft, stick or rod, in reference to its crystal shape.

Crystallizing in the tetragonal system of symmetry, scapolite crystals favour elongate prisms with squarish cross sections and pyramidal terminations, striated along the length. They range through colourless, pale yellow, pale pink and violet, reddish brown and grey. Purple and yellow are the most popular as faceted gemstones. Scapolite is pleochroic showing two distinct colours: pink stones are pale pink/colourless, yellow material pale yellow/colourless and violet stones have strong violet/dark blue. Scapolite is known for its often strong fluorescence under UV light in a range of colours including yellow, red, pink, blue and white.

▲ A 76 ct clear step cut scapolite from Brazil.

▼ A 3.04 ct transparent cabochon with sharp cat's eye, from Myanmar.

Scapolite can occur in large clear crystals, and faceted gems are known of over 100 ct for colourless and over 70 ct for yellows. Many faceted gemstones are eye clean. Inclusions may be needles, hollow tubes and solid crystals of other minerals. Scapolite may be chatoyant from numerous parallel inclusions producing an excellent cat's eye effect in pink, brown, grey,

blue and white stones, when cut en cabochon. Rarely a four-rayed star is produced. Scapolite with inclusions that exhibit iridescence is also known.

This mineral forms in metamorphic rocks, such as marbles, skarns and gneiss. Meionite is more prevalent in rocks of higher metamorphic grade. Scapolite may also occur in igneous rocks.

Although a reasonably common mineral, scapolite is rare in gem quality. The main sources include Mogok in Myanmar, producing cat's eyes in a range of colours including pinks, and faceted stones of colourless, yellow, pink and violet. Madagascar produces clear yellow stones, Brazil produces colourless to golden yellow faceted stones, Kenya purplish pink stones, Tanzania orangey to pinkish cat's eyes and yellow stones, and a purple variety sold under the trade name petschite. Afghanistan and Tajikistan produce purples, and Canada is known for massive opaque yellow or grey material intergrown with diopside, fashioned into attractive cabochons with strong fluorescence.

The moderate hardness and good cleavage mean scapolite requires careful handling, avoiding sharp knocks. If set in jewellery, protective settings should be used and scapolite is not durable enough for everyday wear. Scapolite is generally not treated. Yellow or colourless material may be irradiated to produce purple, however the colour is not stable unlike naturally coloured purple scapolite. Heat treatment may also improve colour and is not detectable.

Sometimes scapolite may be mistaken for quartz varieties rock crystal, citrine and amethyst because of their similar colours, and overlapping values for specific gravity, refractive index and birefringence. Pink stones are similar to kunzite and topaz. Scapolite may be distinguished by its fluorescence and lower hardness. It is also optically negative (quartz is optically positive) and may have higher birefringence.

Sillimanite

Composition Al$_2$SiO$_5$ aluminium silicate
Crystal system orthorhombic **Hardness**
6–7.5 **Cleavage** perfect **Fracture** uneven
Lustre vitreous **SG** 3.20–3.31 **RI** 1.652–1.685
Birefringence 0.014–0.023 **Optical nature**
biaxial + **Dispersion** 0.015

Sillimanite is rare as a gem, and one of the most challenging minerals to facet, but the resulting gemstones are bright with flashes of pleochroic colours, and quite spectacular. The mineral is named after American chemist Benjamin Silliman, and is also known as fibrolite, as it commonly occurs in intertwined masses of long parallel fibrous crystals. This name is still sometimes applied to the transparent faceted gemstones.

Sillimanite is a common metamorphic mineral, found in high-grade metamorphic gneiss, schist and granulites, and rarely in pegmatites. It is the high pressure, high temperature polymorph of Al$_2$SiO$_5$, sharing the same composition as andalusite and kyanite, but forming under different conditions with a different crystal structure. Sillimanite crystallizes in the orthorhombic crystal system. Dominantly fibrous or massive, on rare occasions it occurs as elongate prisms with poorly formed terminations. Most gem rough is found as waterworn crystals in alluvial gem gravels.

Violet blue is the most sought after colour, more commonly found as grey, greyish blue, greenish or brown. Fibrous crystals tend to be colourless. Sillimanite ranges from opaque to gem quality transparent, but this is rare. It is strongly trichroic. Green stones exhibit yellowish green/dark green/blue. Violet gemstones are the most spectacular, cut to show the violet colour with flashes of the yellow and brown pleochroism. In the rare occurrence of sizable gem-quality crystals, they create magnificent cut gemstones.

▲ A fine 19.85 ct violet-blue faceted sillimanite showing flashes of brown pleochroic colour, from Mogok, Myanmar.

▶ A waterworn gem-quality crystal of sillimanite from Mogok, Myanmar.

While sillimanite is widespread as a metamorphic mineral, gem-quality material is found in only a few locations: Ratnapura in Sri Lanka (green to grey), Odisha in India (colourless to greenish), and Mogok in Myanmar (the finest violet blue). Chatoyant material has been found in Sri Lanka, India, Madagascar and Tanzania, usually opaque grey-green to dark brown with fibrous inclusions creating a sharp cat's eye effect when cut en cabochon.

Sillimanite is incredibly difficult to cut and polish due to its perfect cleavage and splintery fracture. Cut gemstones tend to be rectangular cushion or oval shapes, to take advantage of the elongate crystal form and give protection to corners. Care should be taken when wearing and protective settings are recommended. Violet blue sillimanite may be confused with iolite due to similar yellow pleochroic colours.

Sodalite

Composition $Na_4(Al_3Si_3)O_{12}Cl$ sodium aluminium silicate **Crystal system** isometric **Hardness** 5.5–6 **Cleavage** poor **Fracture** uneven, conchoidal **Lustre** vitreous, greasy **SG** 2.15–2.40 **RI** 1.478–1.488 **Birefringence** none **Optical nature** isotropic **Dispersion** 0.018

▲ Massive sodalite with white calcite veining from Swartbooisdrif, Kunene, Namibia.

▶ A cabochon of sodalite with calcite veining.

Sodalite is well known for its rich royal blues used in cabochons and carvings. It frequently has a patchy or mottled appearance due to white calcite veining. The name is derived from its sodium content. The rarely seen variety hackmanite has the interesting property of tenebrescence, the reversable ability to change colour when exposed to light.

This gem is primarily found as opaque massive material in veins in nepheline syenites and other related igneous rocks. Occasionally translucent to transparent crystals occur, favouring a dodecahedral form due to their isometric crystal structure. Although relatively rare, sodalite is found in many locations worldwide. Canada has several important sources including Bancroft in Ontario, where the find of large deposits of quality material elevated it to a decorative stone. Other locations include Brazil, India, Namibia, Italy and Russia.

The finest sodalite of opaque uniform blue is used for cabochons and beads in jewellery. Mottled material, often showing a range of blues with white streaks, is used as a decorative stone for inlays and carvings. Gem-quality crystals may be faceted into beautiful stones. As well as blue, sodalite may occur in grey, yellow, green and pink. Sodalite fluoresces orange under longwave and shortwave UV light, and may show phosphorescence.

Sodalite is tough, with poor cleavage in six directions, appearing as cracks throughout the stone. When polished it has a vitreous to greasy lustre. It is only moderately hard at 5.5–6, so

can be scratched, and protective settings are recommended, especially for rings.

Hackmanite is the sulphur-bearing variety of sodalite, which shows tenebrescence. It occurs as opaque to translucent crystals of grey, pale pink or green. The colour becomes more saturated by exposure to UV light, turning violet, deeper pink or green. The colour will reverse, fading back to paler or whiteish shades, by exposure to white light such as daylight. The presence of sulphur is thought to be the cause. Hackmanite is found in Canada and Greenland, and lesser sources include Afghanistan and Myanmar. It is sought by collectors as mineral specimens.

Sodalite is similar to the intense blue of lapis lazuli and its constituent mineral lazurite, and is often a more affordable substitute. It can be distinguished from lapis lazuli by the incipient cleavage cracks and lack of pyrite crystal inclusions. Synthetic sodalite is known.

Spinel

Composition MgAl$_2$O$_4$ magnesium aluminium oxide **Crystal system** isometric **Hardness** 8 **Cleavage** none **Fracture** uneven, conchoidal **Lustre** vitreous **SG** 3.5–4.1 **RI** 1.711–1.742 (pure 1.718) **Birefringence** none **Optical nature** isotropic **Dispersion** 0.020

◀ Octahedral crystals of spinel from the ruby mines in Mogok, Myanmar.

Spinel is fast gaining popularity as a sought-after gemstone. Its wide range of vivid and pastel colours, combined with high durability, bright lustre and affordable price, make it an excellent choice for jewellery. Spinel has been known for two and a half centuries, but for many years prior was confused with other gems, such as rubies and sapphires, only recently becoming valued in its own right. A find in 2007 of bright red and pink spinel in Mahenge, Tanzania, is credited with bringing natural spinel to popularity. Unusual colours, such as pastel lilacs and particularly greys, are greatly favoured today. Its name is derived from the Latin word 'spinella' meaning little thorn, due to the pointed nature of its octahedral crystals.

Intriguingly, this gemstone has been called The Great Imposter. Spinel is easy to synthesize in a range of colours and has been used as a simulant since about the 1920s. But for centuries prior to this, spinel was mistaken for other gems of similar colours, particularly bright red rubies. With similar appearance and properties, and found in the same deposits, spinel and ruby were believed to be the same stone. It was not until the 1700s, when scientific knowledge was able to differentiate chemically between red gemstones, that spinel was identified as a distinct mineral. It was later discovered that many large and treasured 'rubies' are actually fine red spinels. This includes two famous stones in the British Crown Jewels – the 170 ct Black Prince's Ruby at the front of the Imperial State Crown, and the 352.5 ct Timur Ruby set in a necklace given to Queen Victoria.

The largest fine red spinel is thought to be in the Iranian Crown Jewels, weighing 500 ct. All these large red spinels are polished rather than faceted, partly because they were found and incorporated in jewellery hundreds of years before modern faceting methods were developed. Historically, these stones were known as balas rubies, named for the ancient kingdom of Balascia, related to modern-day Badakhshan in Afghanistan, bordering Tajikistan. This region is credited as the source for these red spinels.

Spinel is found in both metamorphic and igneous rocks. It forms at high temperatures and many of the deposits are related to large-scale tectonic events with associated metamorphism, such as those in East Africa and Sri Lanka (around 600 to 500 million years ago) and the Himalayas (45 to 5 million years ago). Gem-quality material is frequently found in white marble (metamorphosed dolomite-rich limestones) and surrounding pelitic gneiss, and it is often associated with ruby. Due to its hardness and ability to survive weathering, spinel is a common component of gem gravel placer deposits. It can be transported a great distance and for some deposits the origin rock is unknown. The most important sources are Myanmar (Mogok), Vietnam (Luc Yen), Afghanistan/Tajikistan, Tanzania (Mahenge, Morogoro) and Kenya. Important alluvial deposits include Madagascar, Nigeria, Mozambique, Sri Lanka (Ratnapura) and Tanzania (Tunduru, Umba). Other notable sources are Cambodia, Thailand, Pakistan, Nepal, Brazil, Russia and Australia.

Spinel is a magnesium aluminium oxide, with isometric crystal symmetry. Natural crystals commonly form octahedrons, often perfectly shaped with flat crystal faces and a bright vitreous lustre. Surfaces may have triangular etch pits, or terraced growth markings. Spinel is commonly twinned, so much so that this type of twinning in any isometric mineral is called a spinel law twin. This is a twinned octahedron forming a flat triangular shape with v-shaped angles at the corners between each side.

This gemstone ranges from transparent to opaque. It is not often found in large gem-quality crystals, so most faceted stones are less than 2 ct, rarely over 5 ct. Faceted gemstones, particularly lighter colours, are relatively free of inclusions and appear eye clean. Deeper reds and blues may have more inclusions but, due to their rarity, this is deemed acceptable. Inclusions may be fluid or solid, such as crystals of other minerals, including carbonate minerals, apatite, graphite and rutile needles. Zircon inclusions with halos (circular stress fractures around the zircon) are common in stones from Sri Lanka. The most well-known inclusions are orderly rows of tiny octahedral crystals mirroring the crystal symmetry of the spinel. These are either a spinel group mineral like magnetite, or negative crystals that may have a white or black infill. The crystals form strings along rehealed fracture planes and can look like fingerprints. On very rare occasions rutile needles or other fibrous minerals are present in sufficient quantities to produce chatoyancy, revealing a four- or six-rayed star when cut en cabochon.

Part of spinel's appeal is its variety of colours. Pure spinel is colourless, although this is rarely seen in nature, and trace amounts of iron, chromium, vanadium or cobalt impurities result in different colours. Manganese and zinc may also be present. The impurities cause slight variations in the optical properties and density of the specimen, with specific gravity increasing with iron and zinc content.

Bright red spinel is the most popular and valuable. Pink, particularly hot pink, is also desirable. Bright orangey reds are known as flame spinel and a find in Myanmar in the early 2000s of bright pinkish red are known as jedi spinel due to their 'light-sabre' glow. Red and pink are coloured by chromium, orange and purple by chromium and iron. Yellow, caused by manganese, is the rarest colour.

Blue, violet and rarer green often have greyish undertones and are coloured by iron. Pure grey has become very popular, and iron-rich opaque black spinel is attractive when faceted due to the bright lustre. The variety gahnospinel, known from Sri Lanka, is zinc-rich with a blue-violet colour very similar to sapphire. A rare, very bright saturated blue

▲ Mixed cut 9.83 ct spinel gemstone from Mogok, Myamar with deep colour.

◀ A 11.3 ct step cut colour change spinel from Mogok, Myanmar, appearing blue in daylight.

▶ Under incandescent light, the spinel appears purple.

spinel, coloured by impurities of cobalt, is highly desirable and commands high prices. This cobalt spinel was first found in Sri Lanka, with newer finds in Luc Yen, Vietnam. Colour change spinel is known, including a blue to purple/pink change that may be confused with sapphire. Red, pink, violet, and cobalt blue spinel may fluoresce reddish under longwave UV light.

Its high hardness and lack of cleavage make spinel a perfect choice for jewellery. It is cut for its colour with a range of shapes and faceting styles. Its equant crystal shape lends the stone to rounds, ovals and cushion cuts. Spinel's longevity means it is commonly found in antique jewellery, although it is not always recognised as spinel. It is stable and resistant to heat and chemical attack. Warm soapy water is best for cleaning, however unless the stone has noticeable inclusions, ultrasonic and steam cleaners can be used with caution.

Most spinel on the market is untreated. Rarely it is heated or irradiated to improve colour and clarity and this treatment is stable. Fracture-filling to hide surface-reaching fractures is also known. The lack of treatment is of great appeal especially as spinel's main rivals, ruby and sapphire, are commonly treated to improve colour and clarity.

Synthetic spinel was first created in 1907, produced commercially since the 1920s. Most colourless spinel on the market is synthetic,

and it can be made in any colour inexpensively. Those made by the Verneuil flame-fusion method can be distinguished from natural spinel relatively easily. The refractive index is slightly higher (1.727) than natural spinel (1.718). As spinel is isotropic, it should remain dark (extinct) under the polariscope, however the synthetic stones are often internally strained, with thin shadowy stripes moving across the stone as it is turned. This is known as tabby extinction, due to the resemblance to tabby cat stripes. There are also no natural crystal inclusions.

Spinel may be confused with many other gems of similar colours, particularly ruby, sapphire, iolite, tourmaline, zircon, amethyst and garnet, but may be distinguished by its singly refractive nature and different absorption spectrum.

▼ A vivid blue brilliant cut synthetic spinel, 12.35 ct, grown by flame-fusion and coloured by cobalt.

Spodumene

Composition LiAlSi$_2$O$_6$ lithium aluminium silicate
Crystal system monoclinic **Hardness** 6.5–7
Cleavage perfect **Fracture** uneven, splintery
Lustre vitreous **SG** 3.15–3.24 **RI** 1.651–1.681
Birefringence 0.014–0.027 **Optical nature**
biaxial + **Dispersion** 0.017

▶ Gemmy kunzite crystal with etched surface, from Galiléia mine, Minas Gerais, Brazil.

Transparent spodumene occurs in a range of pastel colours, faceted into attractive gemstones. This relatively common mineral is normally opaque grey, white or yellow, and the name comes from the Greek 'spodumenos' meaning burnt to ashes for its resemblance. Spodumene is an important source of lithium, a critical element for rechargeable batteries used in mobile phones and electric cars. As a gemstone it is best known for its colour varieties – pink kunzite, pale emerald green hiddenite and yellow triphane.

Spodumene is a member of the pyroxene group, and thus part of the monoclinic crystal system. It forms characteristic long rectangular to flattened prismatic crystals, striated along the length, with diagonal or pyramidal terminations. It is often etched by dissolution, with triangular markings on the crystal faces.

It forms in lithium-rich granites and pegmatites. It may grow to enormous size, with a single crystal documented to measure more than 14 m (46 ft) in length. Gem-quality crystals can occur in large sizes, and are highly sought by mineral collectors, especially when still attached to the host rock.

Kunzite is the pale pink to violet gem variety, coloured by manganese. It was named in honour of eminent gemmologist and mineralogist George Frederick Kunz, who first identified it in 1902 from the southern Californian pegmatite mines. Kunz was the chief gemmolgist at Tiffany & Co., and helped to bring this stone to popularity. As well as the USA, kunzite is found in Brazil, Madagascar, Myanmar, Afghanistan and Pakistan. As it forms in

large transparent crystals, gemstones can be huge. The Smithsonian holds an 880 ct faceted gemstone from Brazil.

Hiddenite is a rare variety of highly valued pale to bright emerald green. It was first discovered by WE Hidden in 1879 in Alexander County, North Carolina, USA. The nearby town Hiddenite was also named after Hidden in 1913. Hiddenite is coloured by impurities of chromium. As the crystals are generally smaller than other varieties, most faceted gemstones are under 2 ct. Some believe the name hiddenite should be reserved for the transparent green spodumene from this one location, however new finds in Brazil, Afghanistan and Pakistan are often labelled as such. Others use the name for any green spodumene when coloured by chromium.

Triphane is the rare pale yellow gem-quality variety, however the name triphane was used for many years to describe the mineral spodumene, and has only been applied to the gem in recent times.

Spodumene is strongly pleochroic with the deepest colour always seen down the length

◀ Fancy cut 50.79 ct kunzite from Pala Chief mine, California, USA.

▲ A 36.17 ct octagonal step cut hiddenite, from Brazil.

▶ A 78.25 ct oval step cut triphane, from Brazil.

of the crystal. Pink crystals show violet/pink/colourless and the termination can appear a very deep colour due to the pleochroism. Green crystals show yellow green/green/blue green, and yellow crystals pale yellow/yellow/dark yellow. Gem cutters carefully orientate the rough so the table of the gemstone is perpendicular to the length to give the finest colour. Gemstones are fashioned in a range of shapes and cutting styles. Deeper cut pavilions increase the saturation of the delicate hue. Many gemstones are eye clean. Inclusions may be rehealed fractures, aligned growth tubes and etch features. Spodumene may fluoresce deep orange-red in longwave with a weaker reaction in shortwave UV light.

While spodumene is relatively scratch-resistant, with a hardness of 6.5–7, it is brittle and has perfect cleavage in two directions at nearly 90 degrees, common to the pyroxene group. The cleavage, sensitivity to heat, and the strong pleochroism make it a challenging mineral to cut.

Care should be taken when wearing this gem to avoid knocks, and it is better suited to earring or pendant settings, which offer more protection.

Spodumene may be irradiated or heat treated to improve colour. Heat treatment will remove brownish hues from greenish-brown or brownish crystals, leaving light green or pink respectively. Light pink colours may have the colour darkened by treatment.

The colour of spodumene may fade over time, especially with exposure to strong sunlight or heat. This occurs in both naturally coloured and treated gemstones, with treated colours thought to fade more quickly. Always store away from light and heat.

Kunzite is often confused with the pink beryl variety morganite and pink topaz. Triphane may be confused with yellow beryl variety heliodor and chrysoberyl. Kunzite is imitated by synthetic pink spinel and glass, however both may be distinguished by the lack of double refraction.

Tanzanite

Composition $Ca_2Al_3[Si_2O_7][SiO_4]O(OH)$ calcium aluminium silicate **Crystal system** orthorhombic **Hardness** 6–7 **Cleavage** perfect **Fracture** uneven **Lustre** vitreous **SG** 3.35 **RI** 1.691–1.700 **Birefringence** 0.008–0.009 **Optical nature** biaxial + **Dispersion** 0.019

▶ A prismatic tanzanite crystal with striations.

Rarer than diamonds, tanzanite is a relatively new gemstone coveted for its vivid deep blue to violet colour. It is the blue-violet gem variety of the mineral zoisite, and is only mined commercially in the Merelani Hills of Tanzania. It was introduced to the world by jewellers Tiffany & Co., who gave it the more romantic name of tanzanite to highlight the exclusive location. Its rarity and beauty make it very desirable as a gem.

Tanzanite was first discovered in 1967 as scattered blue fragments on the ground of the Merelani Hills. Initially thought to be sapphire, they were identified as a new gem variety. Tanzanite occurs in an area that has undergone extensive tectonic activity, a geological structure known as the Mozambique Orogenic Zone. This belt contains many of East Africa's gem deposits, and is one of the richest in the world. Tanzanite forms in small pockets within pegmatitic veins in graphite-rich gneiss and is thought to have grown over 585 million years ago. The mines only cover a few square kilometres. They were nationalized in 1971, and in 1990 were split into four blocks awarded to different companies or small-scale miners. Tighter controls have been brought in the past decade to control illegal mining and trading. The two largest known tanzanite crystals were reported in 2020, weighing 9.27 and 5.10 kg (20.44 and 11.24 lbs). Prior to this the largest known was 3.38 kg (7.45 lbs).

Zoisite, of which tanzanite is a variety, is a sorosilicate mineral with calcium and aluminium. It is part of the orthorhombic crystal system,

▼ A heat treated 13.39 ct cushion cut tanzanite with violet to blue saturated colour.

and forms elongate terminated prisms with rectangular cross section and striations down the length. It may form in large gem-quality crystals, highly sought by mineral collectors.

◀ A 3.8 cm (1.5 in) tanzanite crystal displaying the three pleochroic colours when viewed from each side and the top.

As tanzanite is cut for its velvety colour, the strong pleochroism has the greatest influence on the cutting process. Rough is orientated to achieve the best blue violet to violet blue colour, and in some stones the pleochroism causes attractive red flashes. Cutting the rough to obtain the finest colour often sacrifices weight resulting in a smaller gemstone. However these top-quality stones command a higher price per carat, and can rival a fine sapphire.

Tanzanite ranges from transparent to opaque. Few faceted gemstones have inclusions visible to the eye. When present, inclusions are commonly wavy fractures, needles or crystals of graphite. Inclusions lower the value, and fractures also effect durability. On rare occasions parallel, needle-like inclusions produce a chatoyant cat's eye effect when cut en cabochon. Gem-quality rough is faceted into a range of styles and shapes with cushion, oval, trillion (triangular) and round shapes common. Gemstones are generally under 5 ct although larger gemstones up to 20 ct are not uncommon with the occasional gemstone more than 100 ct. Lesser quality material may be polished and used as beads.

It has a hardness of 6–7 so it is susceptible to scratching, as well as being brittle with a perfect cleavage. This lower durability means it is not suitable for everyday wear and should be worn in protective settings. Extra care should be taken when wearing rings. Avoid high temperatures or sudden temperature changes. Clean with warm soapy water, but avoid ultrasonic or steam cleaners.

Some tanzanite crystals are found naturally blue-violet. Most rough, however, is yellowish to grey or brown. Heating by several hundred degrees

Tanzanite ranges from sapphire blue to amethyst purple with pure blue or saturated violet blue the most desirable. The saturation of colour increases with size, with deeper colours seen in stones more than 5 ct, paler under 2 ct. The colour is caused by the presence of vanadium. Gem-quality zoisite can occur in other colours including green, brown, orange and pink. As tanzanite is the blue to violet variety, other colours should be called zoisite, however are often marketed as fancy tanzanite.

Tanzanite is strongly pleochroic showing three colours (trichroic) of blue/reddish-violet/greenish yellow to brownish. These can be seen when viewing the crystal in different directions – blue and violet colours through the body of the crystal, at 90 degrees to each other, and brownish down the length. Most gemstones on the market are heat treated to remove the brown component.

Celsius (not much more than a household oven) removes the brownish component, turning the stone the desirable violet blue. Heat treatment normally occurs after the gemstone has been faceted, and is permanent and stable. The majority of tanzanite on the market has been heat treated, and as it is near impossible to detect, assume heat treated until proven otherwise. As heating converts the brownish pleochroic colour to a blue or violet similar to the other two pleochroic colours, stones that do not show brown pleochroism may indicate heat treatment. Other colours of gem zoisite which lack vanadium are unaffected by heat treatment.

Naturally occurring blue-violet stones were likely heated underground during metamorphic processes. It is thought the first discovered tanzanite crystals were originally brown, weathered out onto the surface and heated by a bushfire passing over them, turning them blue.

Other treatments include a thin cobalt-rich or titanium-rich coating on the pavilion of the faceted gemstone to improve the colour strength, and, rarely, fracture filling to improve the clarity.

No synthetic tanzanite gemstones are known, however tanzanite may be imitated by synthetic forsterite, which has a very similar appearance. Forsterite may be distinguished by its lower refractive index and higher birefringence. Amethyst closely resembles tanzanite, as do other blue gems such as sapphire, spinel and iolite, identified by differing optical properties. Blue glass imitations and composite stones are known such as glass doublets with a tanzanite crown.

▲ Naturally coloured brown and blue-violet crystal fragments. The brown crystals will be heat treated to turn them violet blue.

Tektite

Composition SiO_2 silica + magnesium, iron, aluminium etc **Crystal system** amorphous **Hardness** 5–6.5 **Cleavage** none **Fracture** conchoidal **Lustre** vitreous **SG** 2.30–2.50 **RI** 1.46–1.54 **Birefringence** none **Optical nature** isotropic **Dispersion** none

▼ A dumb-bell shaped Indochinite tektite from Thailand.

Many gem lovers are familiar with the glassy green sculptural teardrops of moldavite, a member of the tektite group. Tektites are a natural glass rather than a mineral, meaning they are amorphous. Glass forms in nature when molten rock cools and solidifies so rapidly the atoms do not have time to arrange themselves into an orderly crystalline structure. The name is thought to come from the Greek 'tēktos' meaning molten.

◄ A spherical Philippinite tektite from the Philippines. Etching along stress fractures has created the grooved pitted surface.

The origin of tektites has been the subject of much debate and although the exact process is not yet known it is generally accepted they form from the impact of extra-terrestrial material into Earth's surface. During impact, the local terrestrial rock is melted and ejected high into the atmosphere where it cools rapidly. The molten material moves fast through the air and solidifies in characteristic aerodynamic forms.

Tektites are a silica-rich glass. Their exact composition is determined by the composition of the rock debris from which they are made. Thus tektites from the same impact have a relatively homogenous composition even when found great distances apart, but have a different composition to tektites from different impacts. The region in which tektites are found is called a strewn field. The size, direction and angle of impact determines where the molten rock 'splashes' on land. For some tektites the impact crater is known, for others not, but it is now understood the strewn field for one impact can be thousands of kilometres long. The size of the tektite decreases with increasing distance from

the impact, from centimetre to millimetre scale. The shape of the tektite is determined by its speed and rotation in the air during solidification, and can include spherical, elongated, button, teardrop, dumb-bell and bowl shapes.

They are typically green through to yellowish or brown, with the presence of iron causing the green and brown colours. Inclusions in tektites are gas bubbles of spherical or elongated shape, and swirl patterns. On occasion inclusions of extra-terrestrial material may be found. The surface of tektites are often etched by weathering and burial, creating grooved ridges along their length, spikes, or pitted textures, and they can make quite spectacular specimens.

Varieties of tektite are named based on where they are discovered. Australites are found in South Australia and Tasmania, and are dark glassy often button-shaped objects. Philippinites, found in the Philippines, tend to be spherical, teardrop or dumb-bell in shape and characteristically have deep grooves from etching. Indochinites are a dark glass from Southeast Asia with a wide variety

▼ Translucent moldavite with flattened shape, ridged surface, and conchoidal fractures on either side.

the ground, and may show a conchoidal fracture typical of glass. Common inclusions are the characteristic swirls and bubbles, and vermiform (worm-like) lechatelierite (the glass equivalent of quartz). The sculptural surface textures are appealing, and moldavite is often used uncut in jewellery such as pendant settings. The best quality is faceted into transparent green gemstones with a vitreous lustre.

This natural glass is relatively soft at 5–6.5, and fragile, so is best worn in protected settings. Round or oval brilliant cuts, and emerald step cuts are common to both highlight the colour and remove fragile corners. Warm soapy water is safe for cleaning but avoid chemicals, ultrasonic and steam cleaners. Prolonged exposure to light, and extreme changes in temperature should also be avoided.

Tektites are most commonly confused with the volcanic glass obsidian. Obsidian tends to be greyer with more inclusions of tiny black minerals, whereas tektites are clear and swirly. Tektites contain negligible water, and a sliver of a tektite held in a flame will melt into a glass drop while a sliver of obsidian will foam due to the boiling of its water content – however don't try this at home!

of sizes and morphologies, the largest recorded being almost 11 kg (24.25 lbs)! These three tektite varieties have very similar compositions and were determined to be from the same impact. Together with billitonites, they form the largest strewn field in the world. The impact crater has not been identified but is thought to be in Indochina or the South China Sea.

Tektites were first recorded in Europe from the Moldau River (locally Vltava River) in the Czech Republic in 1787. These tektites are known as moldavite, derived from the river name, and are the most well-known on the gem market. All central European tektites (found predominantly in the Czech Republic, plus neighbouring Austria and Germany) were determined to have the same meteoritic source, formed around 15 million years ago from the impact that created the Ries Crater in Germany. Moldavite is the most transparent tektite variety and ranges from forest to bluish green, or brown, with bottle-green shades highly desirable. Its morphology ranges from spherical to oval, or flattened with an aerodynamic teardrop shape. The surface is typically ridged or pitted, likely caused by heavy etching while in

▼ Step cut moldavite with rounded corners, with bubble and swirl inclusions.

Libyan desert glass

This yellow to green natural glass is found only in a remote area of the Libyan desert in Egypt, scattered over tens of kilometres. Called desert glass or silica glass, it is formed of almost pure SiO_2, equivalent to quartz but amorphous with no crystalline structure. It has a hardness of 6, a refractive index of 1.462–1.465 and a lower specific gravity than tektite at 2.20–2.22. It is commonly cloudy due to numerous trapped gas bubbles. Natural pieces have polished and grooved surfaces, eroded by the wind-driven desert sand.

The origin of this natural glass remains a mystery. It was thought to have formed by a comet exploding mid-air above the desert, leaving no crater but producing a heat burst high enough to melt the quartz-rich sand. Current research suggests it was formed by a meteorite impact similar to tektites, with the heat and pressure of the impact melting the sand, however the impact crater is yet to be found, possibly eroded away. Libyan desert glass is thought to have formed approximately 28.5 million years ago. It has been known in modern times since 1932, but used as a gem material thousands of years ago. Most famous is a carved winged scarab set in the centre of a breastplate of gold, silver and gemstones, found in Tutankhamun's burial chamber. Today some material is cut and polished, however the low brilliance and dispersion do not create attractive lively stones.

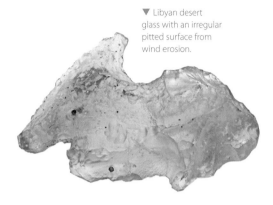

▼ Libyan desert glass with an irregular pitted surface from wind erosion.

▲ A step cut Libyan desert glass gemstone with attractive colour but low brilliance.

◀ Tutankhamun's breastplate featuring a carved scarab made of Libyan desert glass, created over 3,000 years ago.

Topaz

Composition Al$_2$SiO$_4$(F,OH)$_2$ aluminium fluorohydroxylsilicate **Crystal system** orthorhombic **Hardness** 8 **Cleavage** perfect **Fracture** conchoidal **Lustre** vitreous **SG** 3.49–3.57 **RI** 1.606–1.649 **Birefringence** 0.008–0.011 **Optical nature** biaxial + **Dispersion** 0.014

Topaz is a beautiful, bright gem in a wide range of often delicate colours including blue, pink, orange, yellow, brown and colourless. It is one of the most popular faceted gemstones in jewellery for its clarity, high hardness and glassy lustre. As a crystal it forms aesthetic specimens desirable to mineral collectors. Transparent, gem-quality crystals may occur in enormous sizes, allowing for huge cut stones with topaz currently holding the title for the world's largest gem quality faceted gemstone, the Eldorado Topaz.

The name topaz or topazion was historically applied to any yellow or golden gem. The origin of the name is unclear. Many believe it was derived from Topasos/Topazios Island, the ancient Greco-Roman name for Zabargad in the Red Sea (the Greek word 'topazos' meaning to seek, due to the hidden nature of the island). The yellow-green topazion gems mined more than 2,000 years ago are now known as peridot, a different gem to topaz. Another theory about the name is that it comes from the Sanskrit 'tapas' meaning fire. Today the name topaz is still misapplied to yellow gemstones such as quartz topaz (citrine) and oriental topaz (yellow sapphire) so care should be taken to avoid confusion.

Despite the early use of the name, topaz is rarely seen in ancient jewellery with citrine and peridot more prevalent. The mineral called topaz today was possibly not known until much later. The earliest reference to modern topaz is a description by Johann Friedrich Henckel in 1737, of pale yellow topaz crystals discovered in the previous decade at the famous Schneckenstein mine in Germany. The use of topaz has become widespread in the past century, as a gem in its own right, and masquerading as colourless diamond, yellow citrine and blue aquamarine.

The only official variety is imperial topaz, first sourced from the Ural Mountains in Russia in the mid-1800s. This pink topaz was coveted by the Russian Imperial family, hence the name imperial. Today the name is applied to stones with saturated pink, pinkish orange to red-orange colours like the sunset. Other trade names exist including sherry topaz for light brownish-pink to orange-brown.

Topaz is an aluminium fluorohydroxylsilicate, with variable fluorine (F) and hydroxyl (OH). The ratio

▶ An imperial topaz crystal of exceptional colour, clarity and size, with a flawless 96.65 ct step cut gemstone. Both from Ouro Preto, Minas Gerais, Brazil.

of F/OH affects its properties giving a continuous variation in the refractive index and specific gravity. Topaz crystallizes in the orthorhombic crystal system, forming elongate prismatic crystals with lozenge-shaped or rhombic cross sections, capped by flat or pyramidal terminations. The crystals are often sharply formed and very attractive. Striations are common along the length of crystal, and they are occasionally etched.

▶ Sharply formed topaz crystal with natural pale blue colour from Urulga River, Nerchinsk, Russia.

This mineral is found in a wide variety of silica-rich geological environments. It is most common in igneous rocks including granite and associated pegmatites and greisen, and extrusive volcanic rhyolite – it crystallizes in the late stages in veins or cavities from liquids/melts or vapours. The majority of gem topaz comes from pegmatite. Topaz also forms in hydrothermal veins or through metamorphism. Due to its high hardness, it is common as waterworn pebbles in alluvial deposits.

▼ A waterworn topaz crystal from from Aberdeenshire, Scotland.

Well-formed crystals can be more than one metre across, with the largest rough crystal touted at 10 m (33 ft) and 350 tonnes (2,200 lbs) from São Domingos mine in Brazil. The Smithsonian Institution houses two gem quality crystals of 50.4 kg (111 lbs) and 31.8 kg (70 lbs) from Brazil. Blue crystals weighing around 100 kg (220 lbs) are reported from the Ukraine, a gem topaz crystal weighing 117 kg (258 lbs) is in the collection of the Natural History Museum Vienna, and the American Museum of Natural History has the largest topaz crystal on display weighing a whopping 270.3 kg (596 lbs), both hailing from the Fazenda do Funil pegmatite, Santa Maria de Itabira, Minas Gerais, Brazil.

Large gem-quality crystals allow the production of huge gemstones. The largest faceted topaz is the Eldorado Topaz weighing 31,000 ct (6.2 kg, 13.7 lbs), also thought to be the largest gem-quality faceted gemstone in the world. This yellowish brown gem was cut from a 36 kg (79 lbs) crystal found in 1984 in Minas Gerais, Brazil.

Topaz is found in many countries with Brazil the most significant producer. Other sources include Pakistan, Sri Lanka, Russia, Ukraine, Nigeria, Australia, the USA, Mexico, Japan, Namibia, Zimbabwe, Madagascar, China and others. The most important sedimentary gem deposits are in Brazil, Sri Lanka, Australia and Myanmar. Topaz is known from Northern Ireland, Scotland and England, but these are sought for the mineral collector market.

Imperial topaz is known from two locations – Russia and Brazil. Pink stones were discovered in 1853 in the Kamenka River Valley in the Ural Mountains in Russia. Imperial topaz has been known in the Ouro Prêto region of Brazil since the mid 1700s, and the main source today is Vermelhão Mine and the Capão do Lana Mine, famous for reddish-orange stones, with pink, lilac and red considered the most precious. Crystals are not large, and gem quality not common, so 20 ct gemstones are considered large, and over 50 ct rare. The Katlang area in Pakistan provides exceptional pink to violet transparent crystals, however cut gemstones rarely exceed 5 ct due to the small crystal size.

Topaz occurs in a range of pleasing colours, and crystals may have zoning of more than one hue. Pure topaz is colourless and commonly found.

▲ This 58.48 ct step cut topaz from Mogok, Myanmar was pale pinkish-brown, but has faded to colourless.

Blue is the most familiar colour of topaz. Natural blue topaz is actually quite rare, and almost all blue gemstones available on the market have been produced by irradiation and heat treatment. The colours are quite different to the natural paler blue. Swiss blue is the trade term for sky blue colour, and London blue for darker blues.

Yellow to brown topaz is very common, with brownish-yellow to orangey-brown termed sherry. Pale brown colours, such as those from Utah in the USA, Japan and Ukraine, are not stable and fade to colourless with prolonged exposure to light. The highly valued saturated pinkish orange to red-orange of imperial topaz does not normally include these browner hues.

Pink is the most popular colour with a saturated purplish-pink particularly sought after. Natural pink topaz is quite rare and most pinks on the market today are heat treated gemstones from Brazil. Red topaz is rare and highly valued. Natural green topaz is also rare, with dark green topaz produced by treatment.

The cause of colour is varied with pink, red and purple caused by impurities of chromium, and with blue, yellow and brown caused by defects in the crystal structure, possibly by natural irradiation. Orange is a combination of pink and yellow colours caused by chromium together with structural defects. Topaz is trichroic, exhibiting three pleochroic

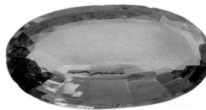

▼ A 40.43 ct pink topaz, faceted in an oval step cut for use in a brooch.

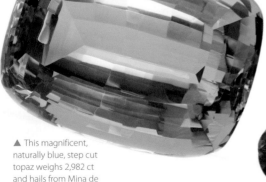

▲ This magnificent, naturally blue, step cut topaz weighs 2,982 ct and hails from Mina de Francisco, Teófilo Otoni, Minas Gerais, Brazil.

colours determined by the body colour – two are distinct and one is weaker. Different colours may show a weak fluorescence.

Topaz is very resistant to scratching, measuring 8 on the Mohs scale of hardness. However, due to its perfect cleavage it has lower toughness, and requires some special care. The cleavage occurs at 90 degrees to the length of the crystal, known as basal cleavage, and is a diagnostic feature. It may be seen on crystals as a flat base with tiny stepped wavy markings. Internally it presents as parallel planes of incipient cleavage, often partially rehealed with trapped fluids creating mirror-like or iridescent planes. The perfect cleavage is a challenge for cutters, and the rough is orientated so that facets are not parallel to the cleavage plane, also reducing the risk of later breakage. Topaz also has a conchoidal fracture.

Most faceted gemstones are transparent and inclusion free. For the rarer, smaller pink and imperial colour gemstones, the value is retained when included. When present, fluid inclusions are characteristic due to the liquid-rich environment in which topaz grows. Prevalent are drop-shaped cavities containing two phase three phase or immiscible liquids, cracks, veils and often iridescent rehealed fractures, mineral inclusions of various colours, and of course the indicative incipient cleavage. Long tubular cavities parallel to the crystal length are common in imperial topaz from Brazil. On rare occasions, topaz is chatoyant with a cat's eye effect seen when cut en cabochon, typically caused by included hollow channels.

Due to the high clarity most topaz is faceted. It takes a high polish due to its hardness, and faceted stones are bright. Colourless and very pale colours favour brilliant cuts to show their brilliance and vitreous lustre, and may be used to imitate diamond or sapphire. The elongated shape of the crystal rough lends to oval and pear outlines. Step or scissor cuts are also popular to best show the colours, with emerald cuts common to protect the corners. Topaz is used in all forms of jewellery, but care should be taken to avoid knocks, especially in rings. Topaz is resistant to chemical attack. Washing with soapy water is best, and be sure to avoid ultrasonic and steam cleaners.

Most topaz is heat treated and/or irradiated to improve or change its colour. Blue is most dominant, but a range of other colours may be produced. These treatments are not detectable by gem testing and should be disclosed. Irradiation of colourless topaz to produce blue is so prevalent it can be presumed any blue topaz on the market has been treated. The resulting medium to deep saturated blue colours are not seen in nature. Irradiated blue topaz has been commercially produced since the 1970s, with a variety of reactors (electron, neutron, gamma radiation) producing different blues. The irradiation may be followed by heat treatment at around 200°C (392°F) to remove any brownish hues, leaving pure blues. Irradiation normally occurs after the stones are faceted. In general the colours are stable to light and heat. Irradiation can also create green topaz, however the colour may not be stable. Irradiated gemstones must be tested for residual radioactivity, and some are stored to allow the radioactive elements to decay.

▼ The 'Ostro Topaz' is the largest blue topaz of its kind at 15 cm (6 in) wide and 9,381 ct. The vivid colour is produced through irradiation and heat treatment.

Many pink topaz gems have been heat treated to deepen the pink colour. Heating red-orange and pinkish-orange chromium-bearing imperial topaz to 400–500°C (752–932°F) removes the orange hues leaving a pink to purplish-red colour. Treated pink colours are stable and unlikely to fade.

Topaz may be fully or partially coated with a thin layer to improve its colour. This may have an iridescent or spotty appearance and can be detected by wear at the facet junctions, or peeling of the coating. The rainbow-like 'mystic' topaz is colourless topaz coated with a thin metallic oxide layer (such as titanium oxide) by vapour deposition, which produces colourful iridescence by thin film interference. In older jewellery, pale topaz may have a foil backing in a closed setting to enhance the colour.

Synthetic topaz is known, however as natural topaz is plentiful and inexpensive, synthetic topaz is generally not seen on the market. Imitations of topaz include glass. Many gemstones may be confused with topaz including blue aquamarine, pink kunzite, pale sapphire, apatite, tourmaline, citrine and danburite.

Tourmaline

Composition complex borosilicate **Crystal system** trigonal **Hardness** 7–7.5 **Cleavage** poor **Fracture** conchoidal **Lustre** vitreous **SG** 2.85–3.40 **RI** 1.610–1.778 **Birefringence** 0.014–0.040 **Optical nature** uniaxial - **Dispersion** 0.017

Tourmaline is one of the most beautiful gems, unrivalled for its colour range. Not only does it occur in every colour, it can have multiple colour zones within one crystal. Tourmaline has been used as a gem material since antiquity but was mistaken for other gems, such as emerald and ruby when identified only by its bright colours. It was brought to Europe from Sri Lanka by the Dutch East India Company, where it was recognised in the early 1700s as a new gemstone, and therefore mineral. It was officially named tourmaline by the end of the 1700s. The name is thought to have its origins in the Sinhalese 'toramalli' or 'turamali', meaning gem/stone of mixed colours, for its many hues. As scientific knowledge progressed, the complex and variable chemistry of tourmaline was recognised to comprise a group of closely related mineral species. For gemstones,

however, they tend to be classed by their colour under a range of varietal and trade names, or simply sold as tourmaline.

Tourmaline has trigonal crystal symmetry typically forming in elongate prisms, or at times short squat crystals, with characteristic rounded cross sections of three or six sides. The terminations vary from flat, pyramidal, to complex with many crystal faces, and even a crown of parallel hair-like crystals. Tourmaline is almost always striated along its length. Crystals may be terminated at both ends, commonly with a different form at each end known as hemimorphic, meaning half-form.

It has two interesting properties due to the asymmetric nature of its crystals – it is pyroelectric and piezoelectric. Pyroelectricity is a phenomenon

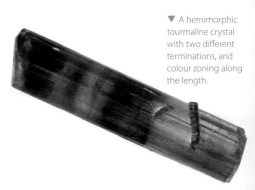

▼ A hemimorphic tourmaline crystal with two different terminations, and colour zoning along the length.

◀ Faceted tourmaline gemstones showing the wide range of colours and cutting styles. The largest gem (top centre) is 83.85 cts.

where heating the crystal generates a positive electrical charge at one end of the prism and a negative charge at the other. This polarity enables a voltage across the crystal that will attract dust or even small pieces of paper This property was first recognised by the Greeks more than 2,000 years ago, and the Dutch used tourmaline in the 1700s to pull the warm ashes out of their pipes, calling them 'aschentrekkers'. Piezoelectricity is the ability to generate an electrical charge when pressure is applied. Tourmaline is used for this property in equipment to detect variations in acceleration and pressure, and was used in WWII submarines to detect explosions. The downside to these phenomena is that tourmaline attracts dust and needs cleaning more often, especially when displayed under jeweller's lights, which gently warm them.

The tourmaline group comprises closely related borosilicates, that is silicate minerals containing the element boron, with the same crystal structure and a range of complex chemical compositions. Many form solid solution series with other members, with elements of similar sizes substituting for each other in the crystal structure. This means tourmaline gemstones are often a blend between species. The varying compositions combined with different chemical impurities cause a range in properties and produce the rainbow of colours. A single tourmaline crystal may have multiple colours due to zones of different compositions or even different species. These crystals are known as bicoloured or parti-coloured.

There are over 37 known species in the tourmaline group, of which five are commonly used as gemstones:

- **Elbaite** – this sodium- and lithium-rich tourmaline is the most abundant in gem quality, occurring in the widest range of colours. It is often parti-coloured with the most prevalent combination being green with red or pink. The majority of gemstones on the market are elbaite. It is named after the island of Elba in Italy.

- **Liddicoatite** – a calcium- and lithium-rich tourmaline that comes in a range of colours, often pink to green and blue, and is frequently parti-coloured. It is relatively uncommon as a faceted gemstone, however large multicolour crystals from Madagascar are often cut and polished into attractive slices. Liddicoatite is named after famous US gemmologist R T Liddicoat.

- **Schorl** – a common sodium- and iron-rich species, opaque black to dark brown, but rarely cut into gemstones due to the opacity. Natural crystals are sometimes used whole as pendants. It may also occur as inclusions within clear quartz, polished into attractive gemstones known as tourmalinated quartz. The name schorl has a history going back more than 600 years, thought to originate from a village known today as Zschorlau in Saxony, Germany, near where black tourmaline was found in a mine.

- **Dravite** – a sodium- and magnesium-rich tourmaline usually dark yellow to brown, black or green. Dravite is named after the river Drave (Drau) in Slovenia along which it was first found.

- **Uvite** – a calcium- and magnesium-rich tourmaline often brown or dark green to black. It is named after the Sri Lankan province Uva where it was discovered.

Tourmaline has been classed by its colour under varietal or trade names for many decades, in part due to the impossibility to determine species without analysis. A tourmaline variety can therefore be more than one species, unusual for minerals! The most well-known are:

- **Rubellite** – one of the most sought after colours of beautiful raspberry pink to red, often with purplish, orangey or brownish tints. Rubellite is more included than other colours and transparent flawless gemstones are highly valued. Due to the rich colour, however, inclusions are deemed acceptable. Red colours may be caused by traces of manganese impurities.

- **Verdelite** – occurs in a verdant range of greens from yellowish to the highly desirable bluish green. Transparent crystals are quite common resulting in brilliant, eye-clean gemstones. Most green tourmaline is strongly pleochroic. The colour is caused by traces of iron.

- **Chrome tourmaline** – a bright green variety coloured by impurities of chromium, vanadium or both. Its saturated green colour can rival emerald and tsavorite garnet, and it is an affordable alternative. It is predominantly sourced from Tanzania.

- **Indicolite** – a range from light to vivid to very dark blue, often with green, violet or greyish hints. Indicolite is less common than rubellite or verdelite. Blues are coloured by impurities of iron.

- **Canary** – bright yellow. Canary tourmaline is found in Malawi. Some are coloured by traces of manganese. Yellow colours may also be created by irradiation or heat

- **Achroite** – a colourless variety of tourmaline, quite rare and not often seen as cut gemstones.

- **Watermelon** – aptly named for its pink centre and thin green 'rind'. It is normally polished into slices to show the colour zoning. Watermelon tourmaline is generally included.

▼ A colour zoned 6.75 ct faceted tourmaline with visible inclusions.

◄ A magnificent 29.4 ct copper-bearing 'Paraíba-type' tourmaline from Mozambique with prized neon colour.

► Paraíba tourmaline from Paraíba, Brazil, set with diamonds in a gold ring.

▼ A spectacular Elbaite specimen from Pederneira Mine, Minas Gerais, Brazil, 24.5 cm (9.6 in) high.

- **Paraíba** – electric green, neon blue or violet. This is the rarest and most valued of all tourmalines, reaching some of the highest prices per carat of all gemstones. It was first found in 1987 near the village of São José da Batalha in the Paraíba state of Brazil, leading to its naming. Unusually, impurities of copper and manganese cause the incredible colour. Copper was not known as a colouring agent in tourmaline before this find. The naturally occurring colours are bluish green, blue, green to pink and purple. Heating may remove the pink and purple, altering the colour to pure neon green, blue or turquoise. Similar copper-included Paraíba-type tourmaline was found in the early 2000s in Mozambique and Nigeria. There is some dispute as to whether these can be called Paraíba if they do not come from Brazil. This variety is mostly elbaite, but some are liddicoatite.

Tourmaline is an abundant mineral found in igneous, metamorphic and sedimentary settings. It is most commonly associated with granitic rocks and pegmatites. The finest gem-quality crystals occur in pegmatite, and with space to grow inside cavities they can form incredible intergrown sculptural specimens. Elbaite and liddicoatite occur in pegmatites, uvite and dravite form in calcium- or magnesium-rich metamorphic rocks such as marbles. Tourmaline may also be found in sedimentary rocks. As a durable mineral it resists weathering, and may be transported far from its host rock, forming rounded pebbles and grains in gem gravel deposits. In recent times it has been recognised that tourmalines, particularly zoned crystals, record a history of the geological conditions in which they grew. This makes them an important mineral in the study of the origin and formation of rocks.

Gem-quality tourmaline is found around the world. Famous locations for colourful crystals include Brazil (predominantly the state of Minas Gerais), Pakistan, Afghanistan, the USA (particularly Maine and southern California), Russia, Sri Lanka, Myanmar, and many African countries including Madagascar, Mozambique, Namibia, Tanzania, Kenya, Nigeria and Malawi. Paraíba tourmaline comes from Brazil and Paraíba-type from Mozambique and Nigeria. Chrome tourmaline is mainly sourced from Tanzania and Kenya, rarely Myanmar.

Tourmaline ranges in transparency and many faceted gemstones are clear with very few inclusions. Green tourmaline is characteristically eye clean. Other colours may have some visible inclusions, while rubellite and watermelon tourmaline are typically included. Tourmaline grows in liquid-rich environments so fluid inclusions such as liquid feathers and liquid-filled rehealed fractures are common. Other characteristic inclusions are wavy irregular fractures orientated across the crystal, needle-like mineral inclusions and thin hair-like hollow growth tubes parallel to the length of the crystal, which may be liquid- or gas-filled. If enough parallel inclusions are present, correctly cut stones may show chatoyancy with a cat's eye effect.

Tourmaline has strong birefringence, and the doubling effect may be seen in larger gemstones with the aid of a hand lens.

Colour has the greatest influence on cutting. Tourmaline has moderate to strong pleochroism, showing two colours (dichroic), best seen in green or brown material. The pleochroic colours are frequently light and dark tones of the same colour (eg light blue/dark blue) but may be two different colours (eg blue/green). One colour is seen through the body of the crystal and a second, always darker colour, down the length. Tourmaline also absorbs the most light down the length of the crystal, so much so that a green crystal may appear almost black down its length. This, combined with the strong pleochroism, influences the orientation in which the gem is cut. The elongate crystal form will then determine the overall shape of the gemstone. If the tourmaline is deep in colour, the gem is cut with the table facet parallel to the length of the crystal to display the lightest colour. This results in a rectangular or elongated gemstone, making the most of the long crystal. If the rough is pale, the gem is cut with the table facet perpendicular to the length to take advantage of the deepest colour, resulting in a more equant round, square or triangular shape. This choice is a trade-off between a higher carat weight stone of lesser colour, or a smaller stone of finer colour. Deep pavilions help to strengthen the colour. More included material is fashioned into cabochons or beads.

Tourmaline crystals with multiple colour zones may change in colour along the length of the crystal or from the centre outwards. The rough is cut to make best use of this, faceted to show the different hues. Crystals with an outwards colour zoning, such as watermelon tourmaline, may be sliced and polished. Madagascar is famed for its crystals with intricate colour zoning reflecting the trigonal structure, and are similarly sliced. Parti-colour tourmaline is a favourite for carving.

Chatoyant stones with multiple parallel inclusions will be cut en cabochon to produce a cat's eye effect. The phenomenon is more diffuse than seen in chrysoberyl due to the larger size of the tube-like inclusions. Cat's eye tourmaline may be blue, pink and green and even parti-coloured.

On rare occasions, tourmaline exhibits a subtle colour change when viewed under different lighting. Chromium-bearing tourmaline may show an unusual phenomenon known as the usambara effect, where the tourmaline will change from deep green to dark red as the thickness of the crystal (and therefore the length of the path taken by light passing through the stone) increases.

Tourmaline is quite durable, which combined with its lovely colours makes it a perfect gem for jewellery. It is used in all forms of jewellery as well as carvings and ornaments. It has a hardness of 7–7.5 and does not scratch or cleave easily. It can, however, be brittle, particularly if heavily included or fractured, and should be protected from knocks and sudden changes in temperature. Tourmaline can be cleaned in warm soapy water. Ultrasonic and steam cleaners are not recommended.

Treatments include clarity enhancement by filling surface-reaching fractures and fissures. Resins or oils are used. Heating and irradiation are often used to alter the colour and these treatments are difficult to detect. It is normally done after the gemstone has been faceted. Heat treatment is usually stable. It lightens dark brown dravite, blue and green stones, and in copper-bearing Paraíba tourmaline removes pinkish hues to create pure neon blue to green. Light coloured stones including some greens can also be deepened by heat. Pale pinks may be irradiated to create hot pink and reds. Canary yellow gemstones are normally treated to reduce orange and brown hues to give a purer colour. Irradiation is not always stable and the resulting colours may fade with exposure to light or heat. Tourmaline may also be coated with thin layers to change or enhance colour.

Synthetic tourmaline has been known since 1991, but as the complex chemistry makes it difficult to synthesize, it is not used for gemstones. Gemstones marketed as synthetic tourmaline are normally imitations of similarly coloured synthetic spinel or cubic zirconia. Other imitations include glass sandwiching a layer of coloured plastic, used to imitate a parti-coloured stone.

▼ Cat's eye tourmaline with parallel inclusions. Top to bottom: 3.90 ct grayish blue oval, 8.88 ct pink oval and 4.23 ct yellowish-green round cabochon.

Turquoise

Composition $CuAl_6(PO_4)_4(OH)_8 \cdot 4H_2O$ hydrated copper phosphate **Crystal system** triclinic (microcrystalline) **Hardness** 5–6 **Cleavage** none **Fracture** conchoidal **Lustre** waxy, dull **SG** 2.31–2.91 (stabilised may be lower) **RI** 1.610–1.650 **Birefringence** 0.040 **Optical nature** biaxial + **Dispersion** none

▼ Nodular turquoise with rich colour from Wadi Maghareh, Egypt.

Turquoise is one of the oldest gemstones, valued for its glorious sky blue to blue-green colours after which the colour turquoise is named. It has been used for adornment and carvings for thousands of years by ancient civilisations in Egypt, Persia and China, and is a ceremonial gemstone for Native American people. The name is derived from the French expression 'pierre turquoise' meaning Turkish stone, as the first material to arrive in Europe made its way from Iran and Egypt via Turkey.

Ancient Egyptian civilisations prized turquoise, set into gold jewellery and carved into scarab amulets. Many turquoise-adorned artefacts have been uncovered in tombs, perhaps the most well known being the gold burial mask of King Tutankhamun inlaid with turquoise, carnelian, lapis lazuli and glass. Ancient Persians wore turquoise as a talisman for protection and victory. They used it as a decorative stone, often engraved, and covered palace domes in turquoise, symbolizing heaven by its sky-blue colour. Native American people used turquoise as a ceremonial gemstone in amulets and jewellery, and attached it to hunting bows in the belief it improved the accuracy of the arrows. The Aztecs favoured turquoise, used for inlays in magnificent ceremonial items such as masks. It was thought they traded with Native Americans for the highly valued gem, but recent research indicates the turquoise was sourced in Mesoamerica. Turquoise had a revival in popularity during the Victorian period, and the traditional turquoise and silver jewellery produced by Navajo and other Native American tribes was actually only introduced in the late 1800s.

Turquoise is a copper phosphate mineral, and copper is the element responsible for the attractive bright blue to green colour. The most desirable is pure, highly saturated blue. Impurities of iron increase the green component.

This gem is only found in a few locations around the world. It is a secondary mineral that forms through sedimentary processes requiring certain conditions – arid environments where acidic groundwater leaches copper, percolating it downwards through the dry earth to react with minerals containing aluminium and phosphate. Turquoise is found in the alteration zones of copper deposits, or as veins in igneous and sedimentary rocks deposited in cracks by the groundwater. It is often found in association with other secondary copper minerals such as chrysocolla and malachite, and silica-rich chalcedony.

The mineral occurs as fine-grained opaque to semi-translucent massive material found in veins, botryoidal layers and clusters, or nodules. Crystals are very rare, small and sometimes transparent, and are very collectible. Finer textured material is harder with low porosity, and can achieve a nice lustre when polished. Most material however is powdery or chalky, and porous. Turquoise often contains interesting brown to black veining of included minerals, such as limonite, or host rock (matrix). Delicate web-like patterns are very desirable and may be known as spiderweb turquoise. The pattern of the veining, together with the colour, hardness and provenance are all used to determine the value of turquoise.

The Sinai Peninsula in Egypt is possibly the earliest source of turquoise, known since at least 3200 BC and highly prized. It is usually blue to greenish blue, often with bright blue spots. The Nishapur district of Iran (formerly known as Persia) is the traditional source for the best colour turquoise, thought to be mined since 2000 BC. It is a rich blue with less veining or markings than other locations, or may have white spots. It is known as Persian blue, however this name is often applied to any pure sky blue turquoise without veining regardless of provenance. The USA is another important source that can rival Iran, found in the southwest states of Nevada, Arizona, New Mexico and Colorado. It has been mined by Native American people since around 200 BC. Other sources are Mexico and China, where it has been mined for millennia.

Turquoise is well known on the gem market and is very popular in silver jewellery, which promotes the attractive colour. Material is fashioned into cabochons and beads, or may be used as a decorative stone for carving and inlays. Sometimes the natural nodular surface is left, creating a tactile gem. Cabochons can be reasonable sizes, or small pieces may be used to mosaic into larger items. Turquoise may be reconstituted, where small pieces are powdered and bonded back together in a resin, sometimes called block turquoise.

The high porosity means turquoise may be damaged by liquids such as water, or through chemical attack by cosmetics and perfumes. Even skin oils can damage or discolour turquoise giving a greenish tint. Wipe with a damp cloth using only water, and do not use ultrasonic or steam cleaners. Care should be taken to prevent it being scratched due to its softer nature. Turquoise may discolour, fade or dehydrate with prolonged exposure to sunlight or heat.

Most turquoise is treated due to its porosity and powdery/chalky nature. It is often coated or impregnated to enhance and darken the colour,

A historical bead of fine Persian turquoise with uniform colour. From the old turquoise mines near Nishapur, Iran.

Polished cabochon of turquoise with limonitic veining from Nevada, USA.

Turquoise from Arizona, USA, impregnated with styrene and alkyd resin to improve colour, lustre and durability.

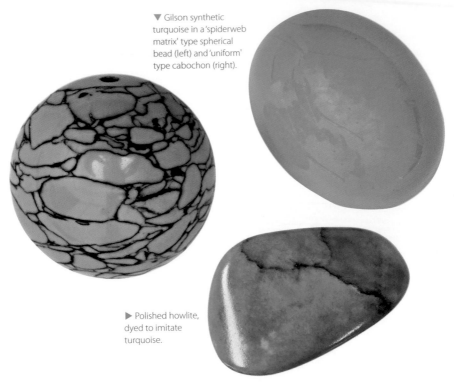

▼ Gilson synthetic turquoise in a 'spiderweb matrix' type spherical bead (left) and 'uniform' type cabochon (right).

▶ Polished howlite, dyed to imitate turquoise.

and to improve the durability, polish and lustre. Wax, oil, epoxy resin or acrylic are used. Treated turquoise is known as stabilised. It may also be dyed to improve the colour, which can be detected by the concentration of colour in cracks. If uncoated, wiping with acetone may reveal dye, but take care as this is a destructive test. Treated turquoise is more affordable than untreated, and all treatments should be disclosed.

Dyed howlite is a common imitation of turquoise. This compact earthy white to grey mineral with grey or black veinlets is very convincing when dyed blue, but can be distinguished by a lower hardness of 3.5. It may be marketed as turquenite. Other simulants include dyed magnesite (lower hardness), plastics, and opaque glass (distinguished by vitreous lustre and conchoidal fracture). Gilson turquoise was created in 1972 and made in uniform colour or spiderweb. Its composition is close to natural turquoise but not an exact match, and so it is a simulant rather than a true synthetic. It can resemble the finest turquoise, but can be identified under the microscope by its texture of tiny spheres.

Turquoise may be confused with other minerals such as chrysocolla (which is softer), and variscite (which has a similar appearance but tends to be greener). Odontolite, also known as occidental turquoise and bone turquoise, was thought to be turquoise that had replaced fossilised bones and teeth, pseudomorphing apatite, the constituent phosphate mineral. However research has shown the turquoise blue gem material is fluorapatite, which may have undergone heat treatment. It is rarely seen, but is an effective imitation of turquoise.

Zircon

Composition ZrSiO$_4$ zirconium silicate **Crystal system** tetragonal **Hardness** 7.5 (6 for low type) **Cleavage** poor **Fracture** conchoidal **Lustre** vitreous, subadamantine **SG** 4.65–4.73 (3.90–4.10 low type) **RI** 1.923–2.024 (1.780–1.850 low type) **Birefringence** 0.036–0.059 (0–0.005 low type) **Optical nature** uniaxial + **Dispersion** 0.039

Not only is zircon beautiful, with a range of colours and diamond-like fire, it is the oldest known mineral in Earth's crust. It was through the study of zircon in 1789 that the element zirconium was discovered. The name is thought to be derived from the Persian word 'zargun' meaning gold coloured. Historically the terms hyacinth or jacinth were applied to gem-quality zircons of yellowish-red, orange or red-brown colour, and the name jargoon is used for colourless to yellowish or smoky zircon. Zircon is most well known as an imitation of diamond, but is an attractive gemstone in its own right.

Zircon is classed as high, intermediate or low determined by its crystal structure. High zircons have normal crystal structure, and their physical and optical properties have the highest values (used as standard) showing the greatest brilliance and fire. These are typically used as faceted gemstones. Low or metamict zircons contain tiny amounts of the elements uranium or thorium, which undergo radioactive decay over time. The radiation damages the crystal structure of the zircon, breaking it down over thousands of years and changing the appearance and properties. The zircon becomes cloudy and greenish, and the lustre may be less bright. The specific gravity reduces from around 4.7 to 3.9 as these heavy elements decay to lighter ones. Intermediate zircons have partial damage from radioactive elements, therefore their properties are between high and low zircons. Most, however, are safe to use in jewellery as studies have shown the radioactivity from zircon is generally insignificant.

▶ Faceted zircon in a range of natural colours, the largest is 27.44 ct (centre). Note almost all have rounded cutting styles to protect the brittle corners.

▲ Zircon on matrix showing typical crystal form and bright lustre.

As the decay of uranium to the lighter element lead occurs at a constant rate, this allows zircon to be dated. This geological clock, combined with zircon's durability and survival through geological processes, means it is a key mineral in geochronology. Scientists have dated zircon crystals contained in rocks at Jack Hills in Western Australia to be 4.4 billion years old (currently the oldest known minerals originating on Earth).

Zircon has a tetragonal crystal structure, often forming rectangular crystals with a square cross section and pyramidal terminations on both ends. Doubly terminated short crystals may appear octahedral. Large crystals are rare, more frequently occurring as small grains in igneous, pegmatitic and metamorphic rocks as an accessory mineral. Due to its resistance to weathering, and high specific gravity, zircon weathers out and accumulates in sedimentary rocks including alluvial placer deposits or heavy mineral sands. Most gem-quality zircon is found as pebbles within these gem gravels. Important sources are Sri Lanka, Cambodia, Thailand, Vietnam and Myanmar. Other notable sources include Nigeria,

Australia, France, China, Madagascar, Mozambique and Tanzania. Low type zircon comes mostly from Sri Lanka. Zircon is also a common inclusion in many other gem minerals forming tiny crystals with a halo, caused by the radioactive decay slightly damaging the host gem.

While zircon crystals are typically a reddish-brown colour, faceted zircon comes in a kaleidoscope, from colourless to autumnal yellow, orange, brown, red, green, striking sky blue, even purple. Blue, red and colourless zircon is actually rare, and most on the market is heat treated. Blues are by far the most popular, then lovely saturated greens and reds. Colourless and golden yellow are also frequently seen. Pleochroism is distinct in blue, red or brown zircon with blue/near colourless in blue heated stones. Zircon may fluoresce under both shortwave and longwave UV light, some with strong reaction. Zircon will often show a distinctive absorption spectrum due to the presence of uranium, some with more than 40 strong lines and narrow bands. However, others may only have a weak spectrum or not show a spectrum at all.

Faceted zircon is normally eye clean. Inclusions lower the value and may be solid crystals or needle-like inclusions, rehealed fractures, fluid inclusions, and lamellar twinning or zoning. Zircon can sometimes have a misty look, especially if becoming metamict. Rarely zircon will exhibit a chatoyant cat's eye effect.

Zircon has a high refractive index and high dispersion, close to that of diamond, which gives it a very bright lustre, brilliance and intense fire. It also has high birefringence with a strong doubling effect easily seen with the eye. The doubling also serves to increase the fire.

It is a durable stone, with a hardness of 7.5 and indistinct cleavage, but it is brittle and will chip easily – scratch-free facets with nibbled edges is a good identification of zircon in older jewellery. Heat treatment increases its brittleness.

▲ Metamict green zircon with internal misty appearance.

has been removed, may produce a dark sky blue or colourless stone. Heating in an oxygen-rich environment by passing air through the heating chamber may produce golden yellow, red and also colourless. Temperatures up to around 1,100°C (2,012°F) are used. The colour may be unstable, particularly in lighter blues, fading back to the original colour. Longwave UV light or heat may also modify or fade colour, so care should be taken to avoid prolonged exposure to sunlight. Some gems can be reheated to return the colour. Zircons from different locations react differently to the heat treatment (due to their varying chemistry), those from Cambodia for instance are well known for their resultant darker blues. Heating metamict zircon to high temperatures can help to heal its crystal structure and improve its appearance, changing low zircon to high type.

Zircon is typically cut in brilliant styles to maximise the brilliance and fire. Shapes without corners, such as round or oval, provide more protection to the stone. Octagonal or emerald-step cuts are used for darker gems to best show the colour. Zircon is a challenge to cut due to its brittleness. The rough is also orientated to minimise the double refraction, otherwise the gemstone can appear blurry from the internal doubling effect. Many gemstones on the market are under 5 ct, although larger are not uncommon with blues up to 10 ct, and faceted gemstones more than 100 ct known.

This gemstone is used in a range of jewellery, but is best suited to pendant and earring settings due to the brittle nature. Avoid knocks while wearing and store separately to avoid chipping. Zircon is resistant to heat and chemicals. Clean in warm soapy water. Ultrasonic and steam cleaners are not recommended.

Heat treatment is prevalent, particularly to create blue and colourless stones, to the point where it can be assumed all blue zircon sold in the gem trade is heat treated. This treatment is generally not detectable. Heat treatment is normally applied to brown, dark red or greyish stones. Heating in a reducing environment, one in which oxygen

Zircon may be confused with many similarly coloured gemstones such as aquamarine, hessonite garnet, topaz and tourmaline. Its high fire, lustre, strong doubling effect and combination of scratch-free facets with nibbled edges are, however, distinctive. Due to a high specific gravity, it has more heft than equivalently sized stones of other minerals, the reverse meaning faceted gemstones of equivalent carat weight will be smaller than other gem materials.

Due to the similar properties, and high fire and brilliance, colourless zircon was used to imitate diamond, particularly early last century. It can be easily distinguished by eye from the doubling of the back facet edges, as singly refractive diamond does not show doubling.

Zircon is often confused with cubic zirconia, a human-made simulant of diamond composed of zirconium oxide. While they may look similar, with high fire, they are two completely different gem materials, only linked through containing the element zirconium. Cubic zirconia is singly refractive and can be distinguished by the lack of doubling.

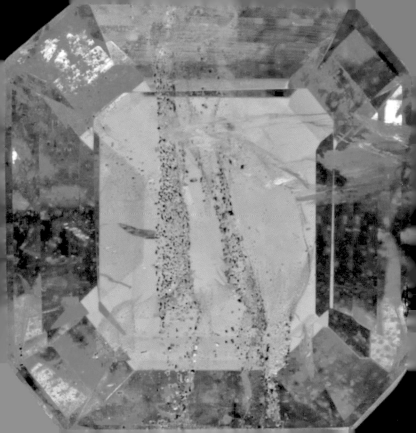

3 Lesser-known gems

While just as beautiful as the more popular gems, the following selection are less well known due to their rarity or low durability. Some, such as benitoite, have a limited supply, while others, like sphalerite, are challenging to fashion and less suitable for adornment. As such, they are often considered collector's gemstones, admired for their splendour.

Benitoite

Composition BaTiSi$_3$O$_9$ barium titanium silicate
Crystal system hexagonal **Hardness** 6–6.5
Cleavage indistinct **Fracture** conchoidal
Lustre vitreous **SG** 3.61–3.68 **RI** 1.754– 1.804
Birefringence 0.017 **Optical nature** uniaxial +
Dispersion 0.039–0.046

Few gem enthusiasts are familiar with this beautiful blue gemstone, which gets its name from San Benito County in California, the USA. It was found here in 1906, and this remains the only source of gem-quality crystals. Benitoite was named the state gemstone of California in 1985, but sadly the deposit is now exhausted, closing in 2005, making this a particularly rare gemstone.

◀ Triangular blue benitoite crystals on matrix from Benitoite Gem Mine, San Benito County, California, USA.

◀ A rare and collectible kornerupine with appealing bright green colour, from Kenya.

▶ Brilliant cut 1.02 ct benitoite gemstone, from Benitoite Gem Mine, San Benito County, California, USA.

▲ Matching benitoite gemstones of 5.95 and 5.88 ct, from Benitoite Gem Mine, San Benito County, CA, USA.

Benitoite is part of the hexagonal crystal system and occurs as distinctive flattish or tabular triangular forms. In San Benito, the crystals occurred in white natrolite veins in grey-blue glaucophane schist. Crystal specimens are sought by collectors for the rareness, attractive colour and shape, and mineral combinations. The location boasts six type specimens, including benitoite.

This gem is typically blue, but may also be colourless, rose pink or orange. The beautiful blue to purplish blue colour is caused by impurities of iron combined with titanium, the same as blue sapphire. Benitoite is strongly pleochroic blue/colourless (dichroic). It has high dispersion, creating fire similar to diamond, although the rainbow flashes are often masked by a deep body colour. Benitoite has high birefringence and a doubling effect is visible in larger gemstones. Inclusions may

be minute crossite fibres, fractures and two-phase inclusions. Benitoite exhibits strong blue fluorescence under shortwave UV light, and is inert under longwave UV light.

Cutting is challenging due to the tabular nature of the crystals, and the orientation of the rough required to show the desirable blue pleochroic colour face up. These factors, combined with the small nature of the gem-quality crystals, means most faceted gemstones are very small, and rarely more than 1 ct. The largest known cut stone is just over 15 ct. Due to the rarity of benitoite, the quality of the cut is sometimes sacrificed to obtain a higher weight gemstone.

Benitoite is moderately hard and has a brittle nature. It is recommended to use protective settings and avoid everyday wear. Due to the rarity benitoite tends to be a collector's stone.

It is not normally treated, although rare colourless gems may be heat treated to a peachy colour. Benitoite may be mistaken for the similar colours of sapphire, tanzanite and iolite, distinguished by its strong dichroism, specific gravity and refractive index.

Calcite

Composition CaCO$_3$ calcium carbonate **Crystal system** trigonal **Hardness** 3 **Cleavage** perfect **Fracture** conchoidal **Lustre** vitreous **SG** 2.65–2.94 **RI** 1 486 – 1.740 **Birefringence** 0.172–0.190 **Optical nature** uniaxial - **Dispersion** 0.008–0.017

◄ Water clear calcite crystals from Gillfoot Park Mine, Cumbria, England.

Calcite is the most abundant carbonate mineral, occurring in a wide variety of colours and holding the crown for the most crystal forms. Although a delicate gem material and a challenge to facet, it can create gemstones with incredible spectral colours.

This mineral crystallizes with trigonal symmetry and there are more than 800 different forms known. Most common are the rhombohedron and scalenohedron, but it may be fibrous, acicular, prismatic, or wafer-thin tabular. Twinning is common. Calcite and aragonite are polymorphs, meaning they have the same chemical composition CaCO$_3$, but different crystal structure, with aragonite crystallizing in orthorhombic symmetry. When microcrystalline they are difficult to distinguish.

It is found in a range of sedimentary rocks such as limestones and, together with aragonite, is a main constituent of the shells of marine organisms. It also occurs in igneous and metamorphic rocks such as marble. Stalactites and other cave formations are made of calcite, precipitating in microcrystalline layers from calcium-rich groundwaters. Calcite is the primary ore of calcium, used in chemicals, drugs, fertilisers and glass. It is mined for lime for cement and is also known as limespar. Calcite-rich rocks such as marble and travertine (a type of limestone) have an attractive appearance and are mined for building stones and ornamental purposes.

Calcite has perfect easy rhombohedral cleavage, that is the cleavage is orientated in the three directions of a rhombohedron – one of its identifying features. It is the mineral used to define the hardness of 3 on the Mohs scale, only just harder

▲ The strong double refraction of calcite seen by the doubling of the cross placed behind a cleaved crystal.

than your fingernail of 2.5. Its low hardness and toughness do not make it suitable for jewellery, thus it is a collector's gemstone.

Icelandic spar is an optically clear variety of calcite, originally from Iceland. It occurs as large colourless transparent crystals that are cleaved into rhombohedrons. It was noted for its strong double refraction in the late 1600s, and this property helped to support the theory of light as waves. This effect is clearly seen with the eye by the doubling of images placed behind a clear crystal. The strong double refraction is utilized in radar, polarizing microscopes and other optical devices. Intriguingly, fossil trilobites have unique compound eyes with lenses of clear calcite crystals.

▲ 1030.5 ct faceted
rainbow calcite, from
Mato Grosso do Sul,
Brazil.

extremely rare due to the difficulty of cutting this very soft mineral with a high risk of cleaving. Most faceted calcite is transparent colourless to yellow, orange or brown, or translucent pink when cobalt-rich.

Rainbow calcite is amongst the most spectacular gemstones. It is cleverly cut from a clear twinned crystal, so that the twin plane is orientated at an angle to the table. Light interacts with the twin plane and this, combined with calcite's dispersion and strong double refraction, multiplies the incredible fire. The best rough comes from New York State, the USA.

Pure calcite is colourless, and impurities of other elements or minerals cause subtle yellow, red (iron or iron oxides), orange, lavender, pink (cobalt), green and blue colours. Calcite is often strongly fluorescent under longwave or shortwave UV light in a variety of colours.

Calcite may be faceted, fashioned into cabochons or carved as ornamental objects. Large gemstones are

Massive material may be dyed to imitate jade, known as Mexican jade. This is often impregnated to improve the hardness and toughness. A banded variety known as limestone onyx or marble onyx is used as a decorative stone, however as true onyx is a variety of quartz, this misleading name should be avoided.

As well as having low durability, calcite is highly reactive to even weak acids, and should be protected from make-up, hairspray, and household cleaners (particularly important if considering a marble worktop or sink!).

Kornerupine

Composition $Mg_3Al_6(Si,Al,B)_5O_{21}(OH)$ complex borosilicate **Crystal system** orthorhombic **Hardness** 6–7 **Cleavage** good **Fracture** conchoidal **Lustre** vitreous **SG** 3.25–3.45 **RI** 1.660–1.690 **Birefringence** 0.010–0.017 **Optical nature** biaxial - **Dispersion** 0.018

Kornerupine is rare as a gemstone and mineral specimen, primarily sought by collectors. It occurs in a range of greens and browns, highly desired when bright emerald green and teal blue. On occasion, gemstones exhibit cat's

▲ A 1.52 ct step cut
kornerupine from Kenya with
exceptional bright green colour
caused by traces of vanadium.

eye phenomenon and very rarely asterism. Kornerupine was named in honour of Danish geologist Andreas Nikolaus Kornerup, who explored Greenland where the mineral was discovered. It is a rare borosilicate mineral, occurring as long prismatic crystals of orthorhombic symmetry. It forms through metamorphism of boron-rich volcanic and sedimentary rocks, however is more often found in gem gravel placer deposits. The main sources are Sri Lanka (including cat's eyes), Myanmar (including cat's eyes and rare star stones), India (including cat's eyes) and Madagascar. Bright green kornerupine occurs in Kwale, Kenya, and blue to bluish green is found in Kenya and Tanzania. Minor sources include Australia and Canada, and rarely Greenland produces gem-grade material.

Kornerupine has vitreous lustre and is normally dark green to brown or golden. The rare bright green colour is caused by the presence of vanadium, and the even rarer blue to bluish green is caused by chromium. Strong trichroism is a distinctive feature, with pleochroic colours varying with body colour and source. Green to brown gemstones from Sri Lanka and Madagascar show brown/greenish/yellowish. Bright green stones from Kenya exhibit green/light green/greenish yellow. Bluish material from Kenya and Tanzania shows green/purplish/bluish. When cleverly cut, more than one pleochroic colour is seen face up, and gemstones from Tanzania with a sea green to violet colour are particularly attractive.

Transparent kornerupine is faceted, while lesser gem-quality material may be fashioned into cabochons. Most gemstones are small, under 3 ct, but up to 20 ct are known. The smaller faceted gemstones tend to be eye clean, while larger ones are more likely to have inclusions of parallel growth tubes and solid crystals of graphite flakes or zircons. Parallel,

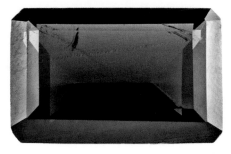

▲ A brownish-green step cut kornerupine of 5.28 ct, from Madagascar.

▼ A green step cut kornerupine of 2.69 ct, from Myanmar.

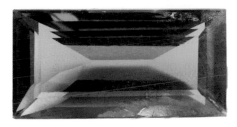

needle-like inclusions of rutile cause the chatoyant cat's eye effect in dark green, brown or yellowish stones.

Kornerupine forms a solid solution series with prismatine, a very similar borosilicate with higher boron content, named for its long prismatic crystals. Confusingly kornerupine is often traded as prismatine even though they are two distinct mineral species. As they are impossible to distinguish without analysis, the name kornerupine is generally applied. However recent research indicates that many assumed kornerupines may actually be prismatine.

With moderate hardness, but good cleavage, care should be taken to avoid knocks. Kornerupine may be mistaken for emerald, andalusite, tourmaline and pale tanzanite due to the similar green, brown or bluish colours, but can be distinguished by the differing pleochroic colours and refractive index.

Painite

Composition $CaZrAl_9O_{15}(BO_3)$ complex borate
Crystal system hexagonal **Hardness** 7.5–8
Cleavage not known **Fracture** conchoidal
Lustre vitreous **SG** 4.00–4.03 **RI** 1.787–1.816
Birefringence 0.028–0.029 **Optical nature**
uniaxial - **Dispersion** not known

Painite is an incredibly rare mineral, at one
time considered the rarest on Earth. It was first
described in 1956 by scientists working at the
Natural History Museum, London, and named
in honour of Arthur Charles Davy Pain, a British
gemmologist and mineralogist who worked in
Myanmar and discovered the first specimen. He
found an unusual deep red crystal in the gem
gravels near Ohngaing, which he sent to London
for identification, where it was found to be a new
mineral. In 1959 a second crystal entered the
Museum's collections, and in 1979 a third was
found and donated to the Gemological Institute
of America. Up until 2001 these were the only
three crystals of painite known to exist. Then, in
the mid-2000s, many more crystals appeared, and
finally the source location was found in Myanmar.
Intriguingly, in late 2007, scientists at the Natural
History Museum, London analyzed a specimen
from Mogok, Myanmar of ruby growing on brown
crystals labelled as tourmaline. They discovered
the brown crystals were in fact painite. This larger
specimen was acquired in 1914 by the Museum,
around 40 years prior to painite's discovery. It had
lain metres away from the two first known crystals,
unidentified for nearly 100 years.

Painite is a borate mineral that contains zirconium.
One of the reasons it is so rare is that the elements
boron and zirconium do not normally occur
together in nature. Painite, along with ruby, forms
in skarns at the contact between leucogranite
and marble, and most is found in the associated
secondary gem gravel deposits. Very rarely it is
discovered still attached to its host rock with ruby
crystals. Myanmar remains its only source, where
several localities are known.

▶ The first (bottom
left) and second known
crystals of painite, with
the ruby on brown
'tourmaline' specimen
(above) which was
discovered to be painite
nearly 100 years after its
acquisition.

With hexagonal symmetry, painite crystals are
elongate prismatic, often with pseudo-
orthorhombic form. They are generally
small, mostly less than 1 cm in
length and often fractured.
High-quality gem material
is rare. The crystals are deep
red to brown, exhibiting red/
pale brownish-orange pleochroic
colours, or pale pink with pale
orangey-pink/colourless pleochroism.
It has a hardness of 7.5 to 8, with vitreous
lustre. Although transparent, painite generally
contains feather-like inclusions, small tabular
crystals and fractures. It is faceted in a range of
shapes, or polished into cabochons. Due to the
small crystal size most faceted gemstones are
less than 1 ct. Painite appears similar to garnet,
zircon and ruby, but is distinguished by several
differing optical properties.

Phenakite

Composition Be_2SiO_4 beryllium silicate
Crystal system trigonal **Hardness** 7.5–8
Cleavage good **Fracture** conchoidal **Lustre**
vitreous, greasy **SG** 2.90–3.05 **RI** 1.650–1.696
Birefringence 0.015–0.019 **Optical nature**
uniaxial + **Dispersion** 0.015

A 19.11 ct brilliant cut phenakite, water clear with high brilliance, from Russia

Phenakite is a relatively rare mineral used occasionally as a gemstone because it creates water-clear, lively stones. Its name, also spelt phenacite, comes from the Greek word for deceiver or cheat, due to its similarity to quartz. Phenakite is of interest to mineral collectors for its attractive transparent crystals. Being rare, the faceted gemstone is considered a collector's stone.

This mineral is a beryllium silicate and crystallises in the trigonal crystal system. It forms short prismatic, tabular, lenticular or rhombohedral crystals. The terminations are commonly a flattened pyramidal shape. It can look very similar to rock crystal, the colourless variety of quartz. Phenakite is often twinned with a penetration twin, which causes a hexagonal prism with a drill bit shaped termination, frequently seen in crystals from Myanmar.

A phenakite crystal formed of penetration twins creating a hexagonal prism with drill-bit termination.

It is found in granite, granitic pegmatite and schist. The best specimens come from the Ural Mountains in Russia, where this mineral was first discovered, associated with the other beryllium minerals emerald and alexandrite. Gem-quality crystals are also found in Brazil, Myanmar, Nigeria, Namibia, Madagascar and the gem gravels of Sri Lanka. Other locations include the USA, Switzerland, Italy, Zimbabwe, Tanzania and Mexico, with large crystals from Norway, although they are not gem quality.

Phenakite is typically colourless but may be pale yellow or pale pink to brown. Coloured stones will show pleochroism from yellow-orange to colourless. Some colours are unstable and may fade over time with exposure to light. Phenakite may exhibit pale blue or green fluorescence under UV light. Gem-quality phenakite ranges from transparent to translucent, and faceted gemstones are often eye clean. Inclusions may be wavy veils of two-phase inclusions and rutile needles. Rarely, numerous fine needle inclusions create a chatoyant cat's eye effect. Colourless transparent crystals are commonly faceted into brilliant cuts, creating bright clear attractive gemstones with a high vitreous to greasy lustre. As crystals tend to be small, most faceted stones are less than 5 ct. The largest known is a 569 ct faceted oval gemstone, cut from a 1,470 ct pebble found in Sri Lanka.

Sometimes phenakite occurs as tiny inclusions in synthetic emerald and can be used to identify the human-made origin. Synthetic phenakite in its own right is known but rarely seen.

Phenakite is most often mistaken for quartz because of its similar hardness, crystal system, and water-clear appearance. It may also be confused with topaz and beryl of similar colours but can be distinguished from all three by the higher refractive index. It may also be mistaken for diamond due to its brilliance, however its low dispersion means it lacks the same fire. Interestingly phenakite will test positive as a diamond on thermal diamond testers. These testers measure how quickly the surface temperature of the gem changes when heat is applied. Diamonds have a very high thermal inertia, so are slow to change, and can be distinguished from simulants with lower thermal inertia such as cubic zirconia. Phenakite has a high thermal inertia similar to diamond. This property also means it feels cold to the touch.

Sinhalite

Composition $MgAlBO_4$ magnesium aluminium borate **Crystal system** orthorhombic **Hardness** 6.5–7 **Cleavage** none **Fracture** conchoidal **Lustre** vitreous **SG** 3.46–3.51 **RI** 1.662–1.712 **Birefringence** 0.035–0.042 **Optical nature** biaxial - **Dispersion** 0.018

▶ A 74.85 ct mixed cut gemstone of sinhalite, almost certainly from Sri Lanka.

This clear honey yellow to brown gemstone was unidentified until 1952, mistaken for olivine (peridot) due to its similar properties. Long thought to be an iron-rich variety, investigations by scientists at the Natural History Museum, in London led to the description and naming of this new mineral. Almost all specimens analyzed during the investigation were faceted gemstones.

Sinhalite's name was chosen from the Sanskrit 'sinhala' meaning Sri Lanka, as this was the only known source until recent times. There, it is found in alluvial gem gravel deposits as waterworn crystals, and Sri Lanka still remains the source of the best quality material. This rare mineral has since been found in gem gravels in Myanmar, as pinkish crystals in Tanzania, in Madagascar, Russia and non gem-quality from the USA.

It is a metamorphic mineral occurring in skarns where boron is present, at the contact between limestone and granite or gneiss. It is orthorhombic, however crystals are very rare, and it normally occurs as granular masses. Gem-quality material is primarily found as alluvial pebbles. Most faceted gemstones are small, however sinhalite can occur in large sizes from Sri Lanka, and cut gemstones of more than 100 ct exist.

Colour ranges from light yellowish brown to dark brown or greenish brown. It is strongly pleochroic pale brown/dark brown/greenish brown. Sinhalite has a vitreous lustre and creates attractive bright clear gemstones. It may be confused with peridot, chrysoberyl, citrine and zircon. It can be distinguished from peridot by its slightly higher refractive index and specific gravity, and in most instances a different optical sign.

Sphalerite

Composition ZnS zinc sulphide **Crystal system** isometric **Hardness** 3.5–4 **Cleavage** perfect **Fracture** uneven, conchoidal **Lustre** adamantine, greasy **SG** 3.90–4.10 **RI** 2.250–2.500 **Birefringence** none **Optical nature** isotropic **Dispersion** 0.156

▶ An unusually large 282.24 ct fiery sphalerite from Santander, Spain.

Sphalerite is the main ore of zinc and not normally of interest to a gem enthusiast. It is commonly opaque brown to black due to incorporated iron content, but on rare occasions pure, or low iron, gem-quality crystals occur. These are faceted into yellow, orange, red and honey brown gemstones noted for their spectacular fire. Rare green crystals occur due to traces of cobalt. The name sphalerite comes from the Greek 'sphaleros' meaning misleading, as dark varieties were commonly mistaken for the lead ore galena due to the high specific gravity and metallic appearance. Sphalerite is also known as zinc blende, black jack, or ruby jack for red varieties.

It has an isometric crystal structure occurring in well-formed crystals of tetrahedral, dodecahedral or complex forms. It may be granular or even globular. Sphalerite forms in hydrothermal veins of sulphide ores, and is found in many locations worldwide. High-quality crystal specimens are sought by collectors from locations including the Tri-state area of the USA and also the north of England. Santander

in Spain is the source of some of the finest gem-quality material, with other locations including Bulgaria, China, Peru and Mexico.

What makes sphalerite so spectacular as a gemstone is the very high dispersion, more than three times that of diamond. Brilliant cuts best show this effect with intense flashes of spectral fire. The dispersion appears highest in light coloured gemstones as darker body colours will mask the fire. Sphalerite has a strong doubling effect easily seen with the eye. Due to its high refractive index it has a very bright adamantine to greasy lustre.

Despite its fire, lustre and attractive colours, sphalerite is considered a collector's stone. It is very soft at 3.5-4, with perfect cleavage in several directions, so is both challenging to cut and too fragile for most jewellery. Synthetic sphalerite is known, but has slightly different optical properties and may be distinguished by its refractive index.

▶ A 4.59 ct brilliant cut sphalerite from Santander, Spain showing high dispersion with spectral colours.

◀ Gemmy sphalerite crystals from Santander, Spain.

Taaffeite

Composition $Mg_3Al_8BeO_{16}$ magnesium aluminium beryllium oxide **Crystal system** hexagonal **Hardness** 8–8.5 **Cleavage** fair **Fracture** conchoidal **Lustre** vitreous **SG** 3.59–3.72 **RI** 1.716–1.730 **Birefringence** 0.004–0.009 **Optical nature** uniaxial - **Dispersion** 0.019

This rare gemstone was first discovered in 1945 by Count Edward Charles Richard Taaffe, a gemmologist in Dublin. It was found in a collection of faceted gemstones, scavenged from old jewellery he purchased from a jeweller. The pale mauve stone was sold as spinel, however Taaffe noticed a distinct doubling effect, thus it could not be the singly refractive spinel. He sent the gemstone for testing at the Diamond, Pearl and Precious Stone Laboratory of the London Chamber of Commerce, with further analysis undertaken at the Natural History Museum in London. It was confirmed in 1951 to be a new mineral. Taaffeite was the first mineral species to be discovered from a faceted gemstone, and was named in honour of Count Taaffe. A second gemstone was uncovered, in 1949, during routine gem testing at the laboratory.

▲ The taaffeite gemstone discovered by Count Taaffe (top, 0.56 ct cushion mixed cut) with the second known gemstone (0.87 ct round mixed cut).

Only four taaffeites were known up until 1967, and it remains one of the rarest gemstones. Its source was unknown for many years, eventually found to be Sri Lanka. Taaffeite forms in carbonate rocks in skarns at the contact to beryllium-bearing granite, but is normally found in alluvial gem gravel deposits. Sri Lanka continues to be the most important source, and a chatoyant purplish brown variety has also been found there. Other sources include Tanzania, China (in limestones), Russia, Myanmar and Madagascar.

Taaffeite has since been determined to be a mineral group, rather than a single species,

distinguished by its iron or magnesium content. The correct species name for this gem is the less romantic magnesiotaaffeite-2N'2S. Its composition is intermediate between spinel and chrysoberyl, with similar properties, so it is unsurprising it was misidentified as spinel. Taaffeite crystallizes in the hexagonal crystal system and is found as colourless, mauve, reddish and bluish crystals. Taaffeite has a hardness of 8–8.5 on the Mohs scale, and is transparent with a vitreous lustre. Inclusions are mineral inclusions or healed fissures.

Titanite

Composition CaTiSiO₅ calcium titanium silicate
Crystal system monoclinic **Hardness** 5–5.5
Cleavage good **Fracture** conchoidal **Lustre**
adamantine **SG** 3.45–3.60 **RI** 1.843–2.110
Birefringence 0.100–0.192 **Optical nature**
biaxial + **Dispersion** 0.051

Titanite, more commonly termed sphene on the gem market, is best known for its strong dispersion, with more fire than diamond. Its rarity in gem quality and low hardness, however, limit its use in jewellery, and it is considered a collector's stone. Titanite is a calcium titanium silicate mineral, named for its titanium content. The popular synonym sphene comes from the Greek 'sphenos' meaning wedge, in reference to the elongate flattened shape of the crystals.

It is a common accessory mineral in felsic igneous rocks and associated pegmatites, as well as metamorphic gneiss, schist and skarn. It is used as a minor ore of titanium. It is found in Pakistan, the Alps of Italy and Switzerland, Brazil, Myanmar, India, Madagascar, Russia, Sri Lanka, Canada, Australia and the USA. Titanite crystallizes in the monoclinic system of symmetry, with wedge-shaped crystals commonly in v-shaped twins. It may also be granular or massive.

This mineral has wonderful autumnal hues of green, yellow-green, yellow, orange, brown or black. Traces of iron and aluminium are the cause of colour with low iron content producing greens and yellow, and increasing iron producing browns to black. Emerald-green crystals occur coloured by chromium impurities, and pinkish varieties due to manganese impurities also exist. Titanite exhibits strong trichroism with the three colours determined by the body colour, typically colourless/yellowish with a third colour of green, brown or orange. Due to the iron content titanite doesn't normally fluoresce under UV light. Titanite

▲ A magnificent 98.41 ct faceted titanite from Madagascar, showing high dispersion.

▼ A 1.37 ct titanite gemstone with a 4.3 cm (1.7 in) twinned crystal from Capelinha, Minas Gerais, Brazil.

is generally transparent, but often included by wavy rehealed fractures, twin planes, tiny crystals or needles. Eye-clean gemstones are rare and highly desired. Its high refractive index produces a bright adamantine lustre and the high double refraction means the doubling effect is easily seen with the eye. Titanite is typically faceted in brilliant cuts, making the most of the dispersion to give the greatest fire. This is constrained by the size and shape of the wedge-shaped crystals, so most gemstones are under 2 ct, although over 12 ct are known. It is normally used in pendants or earrings, which afford the soft stone a protective setting. Most titanite is untreated, although it may be heated to produce orange to red colours.

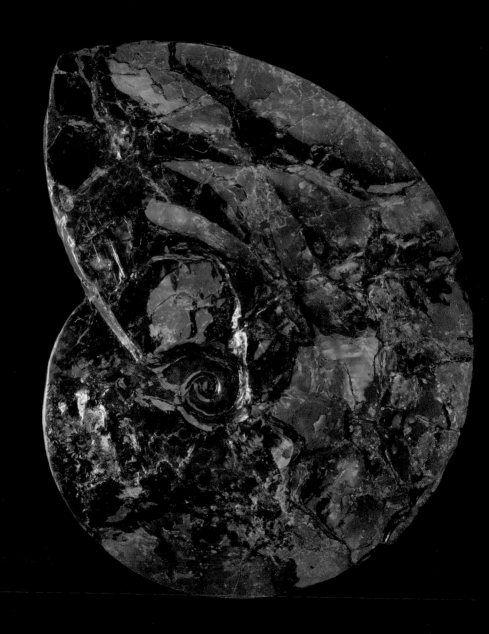

Organic gems

The gems discussed in this chapter are grouped due to their biological origins – they are made by, or from, plants or animals. Some are composed only of organic material, like amber, while others contain inorganic components, such as pearl and coral. These attractive gems have an appealing warmth and beauty.

Amber

Composition C-H-O organic compounds
Crystal system amorphous **Hardness** 2–2.5 **Cleavage** none **Fracture** conchoidal
Lustre resinous **SG** 1.00–1.10 **RI** 1.539–1.545
Birefringence none **Optical nature** isotropic
Dispersion none

▼ An amber pendant set with diamonds, enclosing an insect and spider.

Amber is a fossilized tree resin, treasured for thousands of years for its warm golden colours, ranging from reddish to yellow or white, and on rare occasions blue or green. The name amber is derived from the Arabic 'anbar via Old French 'ambre'. Today amber is prized as much for its science as its beauty, as a time capsule of life gone by. Formed from a sticky resin exuded from trees millions of years ago, it trapped and preserved animal and plant life at that time. More than 6,800 species of insects alone have been identified in amber.

Many trees exude resin as protection against disease or injury. Amber forms when the resin is buried in an oxygen-free environment and slowly fossilizes

◄ Magnificent iridescence is seen in the shell of this fossilized ammonite, known as the gem ammolite.

under the heat and pressure of burial. As it matures it converts to a natural polymer (plastic) and any volatile components (substances that vaporise easily at room temperature) evaporate. The whole process is thought to take millions of years. Younger resins, such as copal, are not amber, even though they may be marketed as 'young amber', as they are not yet fully fossilized and still contain volatiles.

Amber is thought to come from a variety of now extinct trees including conifers and flowering

▼ A pebble of Baltic amber with a polished face revealing the red colour.

▶ Dominican blue amber, approximately 20 million years old, appearing blue due to fluorescence.

species. It is found around the world, but only a few locations are important commercially. The most desired amber is Baltic amber. It formed from a vast forest, accumulating around the Baltic Sea to create the largest deposit on Earth. Most Baltic amber is 34–38 million years old. It is collected along the shoreline, dredged from the Baltic Sea, and mined in open pits. Baltic amber contains 3–8% succinic acid and is sometimes called succinite. Two other sources valued for both gem material and science are the Dominican Republic and Myanmar. Dominican amber is a younger amber of 16–20 million years old, while Burmese amber (also known as burmite) from Myanmar is older, with most 99–100 million years old. All are famed for their well preserved, diverse inclusions of animals and plants. Minor sources include Mexico, Sicily, Romania and Indonesia.

Amber's attractive warm colours range from burnt orange to yellow, brown or white, with reddish colours the most desirable. Rarely it appears blue or green, although green colours may be created by treatment. The colour of a piece of amber may change over time, darkening as it is oxidised by exposure to air. This process can be accelerated by heat and light. Cutting and polishing amber also increases the chance of this occurring as it removes the harder outer crust, and many prefer the deeper 'antique' look as their amber gems age. Blue amber is rare, found mostly in the Dominican Republic, with minor quantities from Indonesia and Mexico.

The blue appearance is caused by greenish blue fluorescence occurring at the surface of the amber, which may actually be of red or yellow body colour. Thus the amber appears blue under daylight which contains some ultraviolet light, with a very strong reaction under UV light. The fluorescence is caused by inclusions of hydrocarbons.

This gem is transparent to cloudy or opaque. Transparent gemstones can be inexpensive but material with interesting inclusions is highly valued. Insects and arachnids are the most common inclusions. Occasionally plant fragments, larger creatures such as lizards, frogs and scorpions and even feathers thought to be from dinosaurs have been found. The amber preserves them in incredible detail, catching them in their last act in three dimensions, not flattened like rock-bound fossils. Other natural inclusions are flow lines, swirls and gas bubbles. Treated amber may contain reflective discoidal internal stress fractures known as sun spangles. Cloudy or opaque material is due to clouds of included gas bubbles, seen under magnification. Amber has a low density with a specific gravity of around 1.08, thus close to that of water (1.00), so it can float in a saturated saltwater solution. Combined with trapped bubble inclusions, it is buoyant in sea water, and is often found washed up on the beach. Amber smells pleasant when rubbed, and it also becomes electrically charged, attracting small pieces of paper by the static electricity (triboelectric)

Amber has been used as a gem material for thousands of years. It is polished into beads, cabochons, and free-form pieces to be set in jewellery. Rarely it may be faceted, more often as faceted beads than cut gemstones. As an organic material it is warm to the touch. Its warmth and lightness make it very comfortable to wear, especially in large pieces, and chunky amber jewellery is popular. A mixture of colours are often used for strings of beads or jewellery set with multiple pieces. Amber is also carved into ornaments and used as a decorative stone. The famous Amber Room at Catherine Palace near St Petersburg, Russia, was created at the start of the 1700s. In its original form its walls were encrusted with more than 6 tonnes (13,227 lbs) of amber, with panels backed in gold leaf. Amber has also been used as a medicine, and oils from burnt amber as a perfume in ancient times.

As a sensitive gem material, amber requires some care. It is brittle and soft with a hardness of 2–2.5, so is easily scratched. Keep amber away from heat as high temperatures can cause it to melt. Heat and strong light can also accelerate the colour to darken, or treated amber, such as dyed, can fade. Clean with warm soapy water and store separately wrapped in a soft cloth. Avoid contact with chemicals such as perfume, hairspray and cleaning products, which can damage it.

Amber is commonly treated by heat and pressure to improve or alter the colour, and enhance the clarity. Heat and pressure treatment is used to clarify cloudy material into clear gem-quality, as well as improving the hardness. Heat and pressure is also used to produce a bright yellowish green colour which does not occur in nature. Amber heated in oxygen will turn a deep reddish 'antique' colour, however this is superficial and may be removed

▲ A ring set with a Baltic amber cabochon enclosing a long legged fly, about 35 million years old.

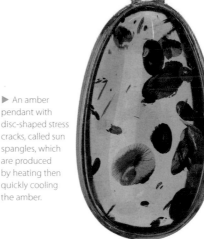

▶ An amber pendant with disc-shaped stress cracks, called sun spangles, which are produced by heating then quickly cooling the amber.

by subsequent polishing. Heating in plant oil of a similar refractive index (like canola) can improve clarity by infilling the bubbles causing cloudiness. Sun spangles are an identifying feature of treatment, caused by a quick drop in temperature and/or pressure, effectively bursting bubble inclusions, creating internal disc-shaped stress fractures. These are often purposefully induced to give an attractive appearance, sometimes named flower amber. amber may also be dyed due to its porosity.

Ambroid is a composite or reconstructed material made by fusing/squeezing together small pieces of amber, using up fragments or scraps from cutting. It can be made in several colours with the appearance of natural amber. It may be distinguished under magnification by patchy translucency and colour, angular boundaries between grains, and gas bubbles elongated by the pressure.

Amber is imitated by younger resin, such as kauri gum and other copal resin, human-made resin (plastics) and glass. Copal and kauri have not yet lost their volatiles and have a pine scent when rubbed firmly. Younger resins tend to craze more readily with a fine network of surface cracks. Plastics have a higher specific gravity and will sink in salt-saturated solutions. Glass may be detected by its cool touch and higher specific

gravity. When viewed between the crossed polars of a polariscope, amber will show interference colours caused by internal strain. The strain is partly caused by the outer surfaces hardening faster than the inner parts, and is also seen around Inclusions such as bubbles and animals, where their final movements are captured as strain within the amber. Plastics and younger resins might show similar interference colours, however this technique may help distinguish heated amber that has released some strain.

There are several destructive tests that can be used with care on inconspicuous areas. A drop of organic solvent, such as acetone, on the surface of amber will not dissolve it, whereas younger resins and some plastics will become soft or sticky. Amber will chip when tested with a sharp knife, while plastics are sectile and will peel. The hot point test, by touching a heated needle point on the amber surface, will yield a piney scent, rather than the acrid scent of burnt plastic.

Inclusions in amber have been forged for hundreds of years, encasing modern animals in plastic, glass or using melted copal. The inclusions often appear too perfect with no signs of a struggle, so beware if it looks too good to be true.

▲ A necklace of imitation amber beads made from celluloid (plastic), phenolic resin, casein (plastic made from milk protein) and glass beads.

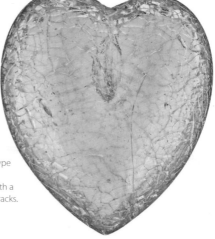

▶ Kauri gum, a type of copal used to imitate amber, with a fine network of cracks.

Ammolite

Composition CaCO$_3$ calcium carbonate + impurities **Crystal system** orthorhombic (microcrystalline) **Hardness** 3.5–4 **Cleavage** none **Fracture** parting along layers **Lustre** vitreous, resinous **SG** 2.60–2.85 **RI** 1.520–1.680 **Birefringence** 0.135–0.155 **Optical nature** n/a **Dispersion** none

Ammolite is the trade name for the highly colourful fossilized shells of two species of ammonite. Ammolite is only found in the Bearpaw Formation, which stretches from Alberta in Canada to Montana in the USA, and only mined from the shales in Alberta. The 71-million-year-old fossils are usually found in ironstone nodules and, apart from being crushed, have hardly altered during burial and fossilization. Ammolite has been known since the early 1900s but was not used as a gem material until 1962. The shell is highly iridescent with a full spectrum of rainbow colour. Red, yellow and green are most characteristic while blue and violet are less common. Similar to modern shells such as abalone, the ammonite shell is composed primarily of aragonite in layers of thin microscopic platelets. The iridescence is caused by the thin film interference effect of light reflecting from the multiple layers of uniform thickness.

Ammolite is categorized in one of two ways: fractured ammolite has a distinctive stained-glass window effect due to the crushed nature of the shell, while sheet ammolite is an unbroken shell of continuous colours. The layers of shell are thin and soft with a hardness of 3.5–4. Ammolite is normally left attached to the shale matrix for stability, or stabilised with resin, then polished. The shell is typically <3 mm (<0.12 in) thick after polishing. Ammolite gems are usually assembled as doublets with a shale backing, or triplets capped by a thin colourless layer of quartz or synthetic spinel. This increases the thickness of the gemstone and the durability. Ammolite is used mostly in pendant and earring settings, which offer more protection, but triplets are durable enough to be worn in rings. Gem material that is not capped should be treated with extra care. Clean with a damp soft cloth, do not use ultrasonic or steam cleaners, and avoid heat and acids such as hairspray and perfume.

▼ A beautiful ammolite (*Placenticeras* sp.) with red to green iridescence, from Alberta, Canada.

Coral

Composition CaCO$_3$ calcium carbonate (black/gold conchiolin) **Crystal system** trigonal (microcrystalline) **Hardness** 3–4 (black/gold 1–2) **Cleavage** none **Fracture** uneven, splintery **Lustre** vitreous, waxy **SG** 2.60–2.70 (black/gold 1.34–1.40) **RI** 1.486–1.690 **Birefringence** 0.160–0.172 **Optical nature** uniaxal - **Dispersion** none

▶ Red coral skeletal branch from the Mediterranean Sea.

Coral is formed from the skeletons of tiny marine animals called polyps, which construct colonies in branching bush- or fan-like structures. Coral has been used in jewellery since ancient times, and was especially popular in Victorian jewellery. Its main source has always been the Mediterranean Sea, and the majority of corals used as gems grow in warm water.

Most coral is composed of calcium carbonate and may be red, pink, white or blue. Black and gold coral is formed of the horny substance conchiolin, an organic material also found in pearls. The most desired coral is the species *Corallium rubrum*, also known as precious coral, with its intense uniform colour of pale pink through to salmon pink and red. When polished it has a very smooth surface, and may show fine parallel markings. Its popularity as a gem has led to the colour 'coral' being named after it. Precious coral is predominantly harvested from the Mediterranean Sea off Italy, France and Spain, and off the coast of north Africa.

Other red *Corallium* species used are found in the western Pacific around Japan, China and Malaysia. Black coral lives off the coast of Australia, Hawaii, Philippines and Indonesia. Gold coral is the rarest colour, found growing off the coast of Hawaii, and is highly desired. Both may show an internal tree-like structure of concentric growth, and gold coral has a finely pitted surface texture. Black and gold coral are softer than precious coral, so can be scratched easily, and may dry out and crack.

The world's population of coral is threatened by overfishing, pollution and environmental changes. The harvest of many corals is restricted and their trade protected under the Convention on International Trade in Endangered Species (CITES). It is important to be aware of the changing regulations and guidelines.

Coral is opaque to translucent with subtle graining from the original skeletons. Colours may be solid or swirled and zoned, with even coloration most desirable. Coral naturally has a dull surface, but takes a high polish with vitreous lustre. It is fashioned as

▼ Coral polished as cabochons and beads, illustrating the variety of colour. Most are precious coral, from Mediterranean and Pacific localities.

◄ Necklace of gold coral beads with shell spacers. The gold coral displays typical pitted surface, and comes from the Hawaiian coasts.

cabochons and rounded beads due to its opacity and softness, and is also left as small polished 'twigs' which are drilled to string as necklaces. It is popular when carved for ornamental use.

As with most organic gemstones, coral is not durable and requires some care. It is soft and prone to scratching, so should be stored separately. Clean by wiping with a damp soft cloth. Corals are easily damaged by heat and acids so avoid hairspray, perfume and cleaning products. The colour is not stable and may fade with prolonged exposure to sunlight. Coral may be bleached to lighten its colour, or bleached then dyed a more desirable hue such as deep red. Dye is detected by colour concentrations in drill holes or surface pits. Bleaching black coral can produce golden colours. Black and gold coral may be coated with resin to boost the colour and improve durability. All treatments should be disclosed.

Coral has been variously imitated by plastics, stained ivory or bone, horn and glass. Plastic imitations may be distinguished by a hot point test yielding an acrid scent. Glass will feel cool to the touch. Both lack the pitted surface, cellular structure or parallel markings. Precious coral may be confused with orange-pink gems such as rhodonite and carnelian, however can be distinguished by lower hardness and warmth to the touch. It may also be confused with conch pearls due to the similar colour and texture.

Ivory

Composition dentine (calcium phosphate + organic material) **Crystal system** none **Hardness** 2–3 **Cleavage** none **Fracture** fibrous **Lustre** resinous, waxy **SG** 1.70–2.00 **RI** 1.520–1.570 **Birefringence** none **Optical nature** n/a **Dispersion** none

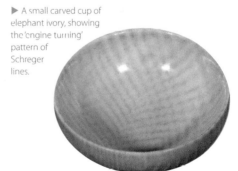

▶ A small carved cup of elephant ivory, showing the 'engine turning' pattern of Schreger lines.

Ivory has been used for thousands of years, as intricate creamy white carvings. It is immediately associated with elephant tusk (a specific type of tooth) to which the term is traditionally applied. However, as the chemical structure of mammal teeth (and tusks) is the same, ivory includes the teeth of other animals such as wild boar, hippopotamus, walrus, sperm whale and narwhal. Fossil mammoth tusk is also included. Tusks are continuously growing teeth that have evolved to project beyond the lips with a specialised function.

This gem is made of dentine comprising approximately 75% of the mineral apatite, and 25% organic material.

The finest ivory is from the African elephant, and is the predominant material used. The large size of the tusks and creamy colour make it ideal for carvings. Indian elephants have smaller tusks, and the ivory is softer and tends to yellow. The high popularity of ivory has led to the risk of extinction for African elephants. An international ban on the trade of ivory was implemented under the Convention on International Trade in Endangered Species (CITES) in 1989 to protect the animals. Sadly, illegal poaching still threatens their survival. Other ivory-producing species have varying protection status. As the rules and guidelines change regularly it is critical to stay up to date, and correct identification of the different types of ivory is crucial. Mammoth ivory is not currently CITES listed as mammoths are already extinct.

Elephant ivory can be identified by its internal structure of micro canals known as dentinal tubules. When viewed in cross section, they form an 'engine turning' pattern of curved crosscutting lines, known as Schreger lines. These lines have angles averaging >100 degrees. Fossilized elephant or mammoth ivory can be distinguished by a smaller average angle <100 degrees between the Schreger lines.

Other ivory will show different identifying features in cross section of concentric or radiating layers around pulp cavities, and different configurations of the micro-canals. Walrus tusks have a darker core of secondary dentine with a porridge-like texture, surrounded by a layer of radiating fibrous dentine, and a thin outer rind of lighter coloured cementum in concentric layers. Hippopotamus tusks are triangular in cross section. The tusks and teeth have a light yellow core consisting of fine concentric layers of regular, even thickness around a small central pulp cavity, and thin outer layers of yellow cementum or thick grooved enamel.

Ivory is primarily carved, used as ornamental objects or for jewellery as carved beads, bracelets and pendants. Carvings often follow the shape of the tooth or tusk. Ivory has also been used for chess pieces, billiard balls, piano keys and buttons. It is a durable material and takes a good polish. Ivory is easily dyed due to its porous structure, often stained to give a darker antique look. Fossil ivory and walrus ivory may be impregnated with wax or polymers to improve durability and lustre.

Ivory has many imitations. Bone has a very similar appearance, but may be distinguished under magnification by its porous nature, with surface pits and grooves. These are Haversian canals through which the network of minute blood vessels ran. Horn is another simulant used, with a structure of parallel fused hairs. Human-made plastics and resins may be distinguished by moulding marks, and lack the bluish white fluoresce shown by ivory and other organics such as bone. Vegetable ivory is the hard nut of certain palm fruit, and its similar creamy colour and hard dense structure makes a good imitation. Sources include the corozo or tagua palm nuts from Central and South America, and doum palm nut from Africa. Under magnification vegetable ivory shows rounded or cylindrical shapes from the plant's cellular structure.

▶ Vegetable ivory is the interior of the nut of certain palm fruit. A sectioned interior portion of Corozo nut from South America (left) and a sectioned doum palm nut from Zimbabwe (right).

Jet

Composition C carbon + impurities **Crystal system** amorphous **Hardness** 2.5–4 **Cleavage** none **Fracture** conchoidal **Lustre** vitreous to dull **SG** 1.19–1.40 **RI** 1.640–1.680 **Birefringence** none **Optical nature** isotropic **Dispersion** none

Jet is an ancient gem used for at least 5,500 years, a favourite of Roman Britain, and hugely popular during the Victorian era for mourning jewellery. The sombre opaque black gem is actually fossilized driftwood, compacted and replaced by carbon, creating a solid material. Jet is worked as carvings, and used as ornaments or jewellery. It takes a high polish and the deep colour is where the term jet black originated. The name jet is derived from the ancient town Gagae, and river Gagates, in southwestern Turkey, where jet was found, via Old French 'jaiet'.

This gem is a hydrocarbon related to coal, formed from ancient coniferous trees. Following death, they became waterlogged and buried in black mud, slowly compacted and fossilized over millions of years. This produced a solid tough material that is found as lumps and masses within layers in shale. Some jet may have a silicified core. There are two types of jet: hard, which is durable and tough, and soft, which is more likely to crack. Research on the oils contained in jet has indicated this difference is caused by the conditions of formation. Hard jet formed under seawater in an oxygen-free environment, while soft jet probably formed under freshwater conditions.

Jet is found in several locations around the world, but is only commercially mined in a few. Whitby in Yorkshire, England, is the most famous source, producing the finest material. It has been mined here for centuries from irregular horizontal seams in shales, but can also be collected from the coastal cliffs or washed up on the beach. Whitby jet is around 180 million years old, and formed predominantly from ancient *Araucaria* trees, similar to the modern monkey puzzle tree. It can have inclusions of isolated quartz grains, from sand trapped in the cracks of the tree. Asturias in Spain produces fine quality jet that is harder but more brittle than Whitby jet. Spanish jet may contain inclusions of brassy pyrite that can make it unstable. Other sources include the USA, France, Germany, Portugal and Turkey but jet from these areas is often of lower lustre and hardness.

As an organic material, jet is warm to the touch and very light, making it comfortable to wear as jewellery. It is tough due to its compactness, but can be brittle with conchoidal fracture. Most jet is smooth but occasionally contains fossils or shows grain or bark textures. Jet naturally has a dull to waxy lustre and may be left matt black, but takes a very high polish becoming reflective in fine material, and was in fact used for mirrors centuries ago. It may be faceted or simply polished into beads or cabochons with a glossy black appeal. Rose cuts are common with faceted domed tops and flat backs. Jet is carved into cameos or ornaments, and may be set in earrings, pendants

▼ Jet from South Beach, Bridlington, Yorkshire, with fossil impressions of ammonites.

▲ Attractively carved jet from Whitby, Yorkshire.

Jet has several identifying properties that can distinguish it. When rubbed on unglazed porcelain, jet leaves a chocolate brown to ginger coloured streak (a streak test). Jet can be intricately carved or faceted with sharp edges, very different to the rounded edges of moulded material. Due to its brittle nature, screws can't be used in jet, so the use of a screw to attach a setting indicates a simulant.

The high popularity of jet in Victorian times led to the use of many imitations including the natural volcanic glass obsidian, bog oak, dyed pressed horn, human-made glass, plastics including Bakelite, and vulcanized rubber known as Vulcanite. Obsidian is distinguished by its cooler touch, glassy conchoidal fracture and higher specific gravity. Bog oak is wood that has been buried in a peat bog and is more textured with a duller lustre than jet. Horn has a layered structure and delamination may be seen under magnification.

Black glass, sometimes called French jet, can be identified by its coolness to touch, internal gas bubbles and swirls, conchoidal fracture, and marks from moulds. It has a higher specific gravity and will not produce a streak. Plastics may be detected by moulding marks, bubbles breaking the surface and lack of streak. Vulcanite, also known as Ebonite, is a common and convincing simulant of similar colour and lustre, with a light brown streak. It is made from natural rubber impregnated with sulphur. Vulcanite is heated and moulded so may show moulding marks, lacks conchoidal fracture, and when firmly rubbed gives a smell of sulphur. The colour typically fades over time with exposure to light from black to khaki brown. Other black gemstones such as schorl tourmaline and onyx may also cause confusion, but can be easily recognised by their coolness to touch.

and brooches. Its popularity soared in the 1800s following Queen Victoria's choice to wear jet as part of her mourning attire after her husband Prince Albert died in 1861. Today jet is still used for ornamental use and jewellery.

Similar to amber, jet is triboelectric and will become electrically charged when rubbed, attracting dust or small bits of paper. For this reason, jet has sometimes been called black amber. Jet, while tough, is soft and easily scratched, so should be stored separately from other jewellery. It can be washed with mild soap and warm water, or wiped gently with a cloth. Do not use ultrasonic or steam cleaners. As an organic material it contains water and therefore can dehydrate and crack. Care should be taken to avoid heat or storage in very dry conditions.

Pearl

Composition CaCO₃ calcium carbonate + conchiolin **Crystal system** orthorhombic (microcrystalline) **Hardness** 2.5–4.5 **Cleavage** none **Fracture** uneven **Lustre** pearly, dull **SG** 2.60–2.85 **RI** 1.52–1.69 **Birefringence** 0.155–0.156 **Optical nature** n/a **Dispersion** none

Pearls have been prized for thousands of years, and are a symbol of wealth and luxury. They are unique as a gem for their warmth, soft lustre and iridescence known as orient. The beautiful rounded shapes come in a range of pastel colours, most recognisably as white, cream or grey. Natural pearls have been used for adornment in jewellery and on clothing for centuries, however their rarity reserved them for the rich and for royalty. The commercial production of cultured pearls from the early 1900s means today everyone can enjoy their elegance.

They are grown by molluscs, and are one of the few gems produced by an animal. The most well known type of pearl is made from a substance called nacre or mother of pearl. It forms within bivalve molluscs such as mussels or oysters, and the sea snail abalone. Other molluscs may produce non-nacreous pearls such as the conch, bailer shell (*Melo melo*), scallop, and the giant clam.

There are less than 20 species of bivalve that produce nacreous pearls. Nacre, or mother of pearl, is secreted by the mollusc and lines the inside of shells. When a microscopic irritant such as a virus or foreign object enters the mollusc's soft body tissue (called mantle tissue), the animal will coat it with layer upon layer of nacre as a protective defence mechanism, forming a pearl. Nacre consists of alternating layers of inorganic calcium carbonate (predominantly as the mineral aragonite) and organic material called conchiolin. The aragonite forms as microscopic platelets of regular size in overlapping parallel layers, bound by the conchiolin.

Pearls occur in different shapes, the most prized being the perfectly rounded sphere. Symmetrical shapes include buttons and drops, and irregular, unusual shapes are called baroque. Seed pearls may be round or baroque and are less than 2 mm (0.08 in) across. Rice crispy pearls look similar to a grain of puffed rice. Blister pearls are those that have grown attached to the inside of the shell, often with a hemispherical shape.

The type of mollusc in which the pearl grows will determine its subtle but complex colour. The overall dominant hue is known as the body colour. It may be neutral (white, grey, black),

▶ A large blister pearl on a *Pinctada maxima* or silver-lipped pearl oyster shell, from the Indo-Pacific Ocean. This oyster produces the large South Sea cultured pearls.

▼ Natural freshwater river pearls from Scotland.

slightly hued (cream, silver, brown) or fancy (pink, lilac, yellow, peach, green, blue). Some show an additional colour of green, pink, purple or blue called an overtone, which overlies and modifies the body colour. Lastly, another dimension may be added by the subtle iridescence of two or more colours shimmering on the surface, known as orient. The cause of the iridescence is complex – a combination of the diffraction of light from the edges of the overlapping aragonite platelets, and the reflection and interference of light from the layering of the platelets.

The soft lustre of a pearl is caused by the way light reflects from the translucent layers. Thicker and better quality nacre will produce a sharper lustre. The water temperature in which they grow will also affect the lustre, and pearls from cooler waters will generally have a better lustre than those from warmer waters. Most pearls will be harvested in winter to obtain the best nacre outer layer. The surface of the pearl will affect both the lustre and iridescence. Smooth, blemish-free surfaces give sharper lustre but are rare. Many pearls will have spots, pits or wrinkles that formed during growth. Some pearls may have lines encircling them, caused by movement during growth, and these are both valued and disliked.

Pearls occur naturally in the wild and are also cultured in farms. Natural pearls are very rare and highly valued, occurring in both saltwater and freshwater. Natural river pearls were coveted from Scotland, but the mussels are a protected species with pearling banned since 1998.

Cultured pearls are farmed around the world using various bivalve molluscs in both saltwater and freshwater. The first cultured pearls were grown in Japan at the end of the 1800s, and commercial production began in the 1920s. The concept of culturing pearls was known as far back as the 1200s in China, where shapes such as buddhas were inserted into pearl shells to be covered with nacre, used for traditional ornamental purposes. The process to culture pearls mimics nature. A small piece of mantle tissue from a donor is inserted into the tissue of the mollusc, with or without a bead (normally a polished sphere of shell) which forms the nucleus. The host mollusc forms a pearl sac around them, secreting layers of nacre. The molluscs are nurtured while the pearls grow, taking six months to several years – the longer the time, the larger the pearl and the thicker the nacre.

Saltwater pearls are produced by pearl oysters farmed in lagoons or atolls. Several species of pearl oyster are used, and the pearls are named after the oyster in which they grow, such as akoya pearls, or the regions in which they grow such as the Tahitian and South Sea pearls.

Akoya pearls are the most popular cultured pearl on the market, and were the first to be commercially farmed. Akoya range between white, cream, golden or pale grey with overtones of pinks and greens. The pink gives a warmth that is more valued, while green has a metallic look. These grow in the species *Pinctada fucata*, or akoya pearl oyster. The oysters are small in size and the pearls range from 2 to 11 mm (0.08 to 0.43 in), normally less than 8 mm (0.31 in) and rare over 10 mm (0.39 in). They are mostly spherical with unblemished surfaces and a sharp lustre, and set the standard for cultured pearls. Japan and China are the main source, but they are also cultured in South Korea and the Mediterranean.

Tahitian pearls are often called black pearls due to their deeper colours, and come in a range of fancy body colours including grey, chocolate brown, green, purple and blue. The overtone colours are stronger against the dark background and may override the body colour. Tahitian pearls form in the oyster *Pinctada margaritifera*, commonly known as the black-lipped pearl oyster. This is a large species, and pearls range from 8 to 17 mm (0.31 to 0.67 in), averaging 10 mm (0.39 in) and rare over 18 mm (0.71 in). They may be spherical or baroque. Tahitian pearls are farmed around the French Polynesian islands including Tahiti and Fiji.

South Sea pearls range from golden to cream, silver or white, with blue, green or pink overtones. They are produced by the species *Pinctada maxima*. These are the largest pearl oysters with pearls predominantly between 8 and 16 mm (0.31 to 0.63 in), up to 20 mm (0.79 in), with thicker nacre. They are also the rarest oysters, so South Sea pearls are the most valuable. The pearls favour oval, button, drop and baroque shapes. There are two varieties of this oyster: the gold-lipped producing golden pearls, farmed in the Philippines and Indonesia, and the silver-lipped producing white, silver and cream pearls, cultured in Australia.

Freshwater pearls are produced by a few species of freshwater mussel, normally cultured in lakes or ponds. They are found in hot and cold climates. Most freshwater cultured pearls use tissue rather than beads (shell) as a nucleus, known as non-nucleated, and therefore consist of more nacre. They are the most common cultured pearl, with affordable prices. Freshwater pearls range in their colours and shapes, often of high quality. The

colours are subtle with lilac, pink, peach, golden and cream. The pearls range from 5 to 15 mm in size (0.2 to 0.59 in). As they don't have a rounded shell nucleus, they tend towards baroque with rice crispy shapes common, however spherical shapes have been achieved in recent times.

The first freshwater pearls were cultured in Lake Biwa, Japan, using Biwa pearl mussels, hence known as Biwa pearls. The USA has a long history culturing freshwater pearls (the only country outside of Asia) but is no longer producing commercially. Today China is the main producer. The cockscomb pearl mussel, *Cristaria plicata*, was originally used, producing unique rice crispy shapes, however in the mid 1990s a switch to the triangle mussel *Hyriopsis cumingii* and mussel hybrids produced more symmetrical, even spherical, higher quality pearls.

Mabé pearls, also called blister pearls, grow against the inside of the mollusc shell rather than in the soft tissue. They are hemispherical with a flat bottom. They can occur naturally or are cultured in saltwater pearl oysters. The most common species

▶ Cultured freshwater non-nucleated pearls from Lake Biwa, Japan, with sharp lustre.

▶ Cultured saltwater pearls, one cut in half to reveal the inner bead (centre).

used is the penguin's wing oyster *Pteria penguin* producing white mabé pearls. The rainbow-lipped oyster *Pteria sterna* grows black mabé pearls, and the South Seas oyster *Pinctada maxima* forms white or gold ones. Abalone *Haliotis iris* produces a mabé pearl of intense iridescence. Cultured mabé pearls are grown by attaching a disk, hemisphere or specially shaped bead to the inside of the shell, which is coated with layers of nacre by the mollusc. The pearls are removed from the shell, the bead removed, filled with a resin, and cemented to a mother of pearl backing. Their shape means they are often used for earrings, rings and buttons.

The value of pearls is determined by their lustre, smoothness of surface, shape and symmetry, colour and size. Thicker nacre produces a better lustre and orient (iridescence), and a more durable pearl. Matching sizes within a set also increase the value.

Most pearls are used in necklace and earring settings, which offer more protection, although rings are common. Pearls are drilled in order to be set or strung. Strings may be of matching sized pearls or graded with larger pearls towards the centre. Pearl necklaces are so loved that different lengths are given official names: choker (41 cm, 16 in) princess (46 cm, 18 in), matinee (56 cm, 22 in), opera (91 cm, 36 in) and rope (132 cm, 52 in).

Natural pearls can be distinguished from cultured ones by their inner structure. Natural pearls are formed of many concentric layers of nacre, sometimes with a small central cavity. Nucleated cultured pearls have a bead with a different banded structure, covered by a thin nacre layer about 0.5 mm thick (0.02 in), thicker in South Sea pearls. Non-nucleated pearls have irregular centres or

cavities with concentric rings of growth. Viewing into the drill holes may reveal the thin layer of nacre on a bead to distinguish cultured. However this is difficult to detect if the beads are strung. X-rays are used in gem laboratories to view the internal structure to determine natural or cultured and the type of pearl. It may also detect imitations.

Pearls may fluoresce, useful for distinguishing a string of natural versus cultured pearls. Natural tend to give a variable reaction while a necklace of cultured treated pearls may show a uniform response. Fluorescent reactions range from blue, green, yellow, red to brown.

Other molluscs may form pearls. Abalone shells can very rarely produce nacreous baroque pearls of high iridescence, commonly horn shaped. The queen conch in the Caribbean produces pink to orange non-nacreous conch pearls. These tend to be oval shaped with a porcelaneous surface and a distinct flame-like pattern. Conch pearls are rare and highly valued. Melo pearls are a rare pearl formed by the bailer shell or *Melo melo* marine snail. They are non-nacreous with a porcelaneous lustre and subtle flame structure. The pearls are relatively large and often round. They are tan to brown and occasionally a soft peach colour, which is highly sought. Giant clams produce white non-nacreous

▶ A choker length (41 cm, 16 in) necklace of cultured saltwater pearls of matching size.

pearls, which are large but not valued as gems. Both conch and melo pearl colours may fade with exposure to light.

Pearls commonly undergo many treatments to alter their colour, and to improve the lustre and durability of these soft gems. Cleaning, polishing and bleaching treatments are accepted as routine practice, and rarely disclosed. Treatments include:

- Polishing to remove rough surfaces and improve lustre, by buffing with beeswax or tumbling in mild abrasives. Pearls may also be skilfully peeled to remove any stained or damaged outer layers of nacre.

- Coatings to enhance lustre and improve durability. Pearls may be polished before coating. Wax and clear polymers are used, however they may peel or wear off over time. Coated pearls will have a smooth surface under magnification, while untreated nacre has a scaly appearance.

- Bleaching with hydrogen peroxide to lighten the colour, whiten, or even out the colour by reducing darker spots.

- Dyeing using organic or inorganic dyes to stain the nacre. This is detected in drilled pearls by concentrations of dye between the bead and nacre, and dye stain on the stringing thread. Dyed pearls may be inert or have dull reactions to UV light.

- Bleaching and dyeing to alter the colour. Nucleated akoya pearls are commonly treated this way, often dyed bright colours. The unnatural hues are a giveaway.

- Treating with silver nitrate solution and exposing to light to create dove grey to black colours. The silver nitrate reacts with light, decomposing into metallic silver as dark deposits within the layers of nacre. The silver may be detected via X-radiography and X ray fluorescence.

- Irradiation by gamma rays to give a greyish body colour, commonly used on akoya pearls. This darkens the bead nucleus and can be identified by grey to black beads under the whiter nacre. The pearls may also be inert to longwave UV light. Gamma irradiation of non-nucleated freshwater pearls darkens the nacre and gives a metallic sheen.

- Heat treatment to alter white pearls to golden. This can be detected by gem laboratories using UV spectrometry.

Imitation pearls are made from glass, enamel or plastic beads with an iridescent coating. A good imitation uses fish scale essence as a coating. Imitation pearls have a smoother surface than real pearls, and the latter feel gritty due to overlapping aragonite platelets. This is sometimes tested by rubbing the pearls against each other. The iridescence is poorer quality and the coating may be seen peeling off the bead. Glass beads may show conchoidal fractures around the drill holes.

Pearls are delicate gems that require care, but if looked after properly will last generations. Pearls are porous, and have very low hardness so are easily scratched. The thin or lower quality nacre of cultured pearls may also chip or crack, however their overall toughness is good. They should be cleaned by gently wiping with a damp soft cloth after each wear, never with steam or ultrasonic cleaners. Store them wrapped in soft cloth, but not in airtight conditions for long periods of time as this may cause them to dry out and deteriorate. Natural oils from our skin help to keep pearls in good condition, so they benefit from being worn. As they consist of calcium carbonate, they are very sensitive to chemicals and acids such as vinegar, cosmetics, perfume and hairspray. As a general rule, put pearls on last and take them off first.

Shell

Composition CaCO₃ calcium carbonate **Crystal system** Orthorhombic (microcrystalline) **Hardness** 2.5–4.5 **Cleavage** none **Fracture** uneven **Lustre** pearly, dull **SG** 2.65–2.87 **RI** 1.53–1.69 **Birefringence** 0.155–0.156 **Optical nature** n/a **Dispersion** none

▶ The abalone (pāua) shell's highly iridescent mother of pearl lining (below) makes beautiful decorative items such as the lid of this trinket box (above right).

Shells have been used for adornment since before the Stone Age. Archaeological evidence suggests they are the oldest form of jewellery dating back more than 100,000 years. Smaller shells may be used whole, drilled for pendants or beads. The mother of pearl (nacre) coating inside the shell of pearl oysters and other molluscs may be used as a gem in its own right.

The abalone shell, also known as paua shell in New Zealand, has a beautiful, highly colourful mother of pearl inner lining with a typically blue-green iridescence. This is a species of sea snail (marine gastropod mollusc) of the *Haliotis* genus. The iridescence is stronger than that seen in other mother of pearl as the layers of nacre are

formed from columnar stacks of aragonite platelets of regular thickness, rather than layers of overlapping platelets. The pāua or rainbow abalone *Haliotis iris* has intense iridescence and can grow up to 18 cm (7 in). The shells are very tough, and are cut and shaped for use in jewellery, buttons and inlays. Pendants and earrings are common, with the shell held in a setting or attached by a drill hole, and small pieces are inlaid to create bracelets. As abalone are popular, both for their shell and to eat, there are restrictions on the size and quantity that may be collected.

Pearl-bearing molluscs such as oysters are also used for the mother of pearl lining of their shells. It has the same soft lustre and iridescence of a pearl. The shell is used for jewellery, buttons, inlays and other decorative purposes.

Shells of different coloured layers may be carved by removing the top surface to contrast with the layer beneath, producing intricate images in relief, known as cameos. The images follow the shape of the shell, and may be cut out or left whole. Pink and white conch (from Jamaica and Madagascar) and brown and white helmet shells are commonly used. Shells are quicker and cheaper to carve than harder stones, and lighter to wear.

Tortoiseshell

Composition keratin (protein) **Crystal system** none **Hardness** 2.5–3 **Cleavage** none **Fracture** smooth, fibrous **Lustre** greasy, resinous **SG** 1.26–1.35 **RI** 1.54–1.56 **Birefringence** none **Optical nature** n/a **Dispersion** none

Tortoiseshell has been used for thousands of years, as a gem material and for ornamental objects. It was very popular from the 1700s to the 1900s. Its warm amber, yellow, reddish and chocolate brown colours have marbled or flame-like patterns, and the highest grade is translucent. The pattern is so distinctive it is used to describe other animals such as tortoiseshell cats.

Tortoiseshell surprisingly does not come from tortoise but primarily from the hawksbill sea turtle, which lives in the tropical waters of the Atlantic, Pacific and Indian Oceans. Nearly hunted to extinction, it is now an endangered species protected internationally since 1977. The trade of tortoiseshell is governed by the Convention on International Trade in Endangered Species (CITES). Trade is banned altogether in some countries, while others may exempt older worked antique material. As the rules and guidelines change regularly it is critical to stay up to date, and correct identification of tortoiseshell and its imitations is key.

The carapace or top part of the turtle shell is used as a gem material. The shell is not shell but made of the protein keratin, similar to hair and nails. Tortoiseshell is moulded by heat and pressure, cut to shape then polished. Several pieces may be fused together to create larger items. Tortoiseshell has been used for glasses frames, combs, boxes, necklaces, bracelets, brooches, ornaments, knife handles and inlays. It is a durable material and, being organic, is warm to touch.

Identification of tortoiseshell is achieved in a few simple ways. When viewed under magnification the darker brown areas are formed by tiny dots of brown or black pigment.

It may fluoresce under UV light with paler areas appearing chalky blue and darker areas remaining dark. A hot point test, undertaken by touching the surface with a hot needle tip, gives the distinctive smell of burnt hair.

Tortoiseshell is frequently imitated by inexpensive plastics, particularly since its ban. These do not show the pigment dots, but have swirls of colour and included gas bubbles. Another common imitation is horn, however it may be distinguished by its structure of parallel fused hairs.

▲ A glasses case made from tortoiseshell, showing the warm translucent colours and distinctive pattern.

Further information

Websites for gemmological and mineralogical information

gem-a.com/ – The Gemmological Association of Great Britain
gembluechart.com/index.html – Gemstone identification
gemdat.org/ – Gemstone database and gemmology information
gemologyproject.com/ – Introduction to gemmology and gemstone database
Gemsdat.be – Compilation of data on properties of gemstones
gemstonemagnetism.com/ – Magnetism in gemstones as an identification tool
gia.edu/ – Gemological Institute of America
mindat.org/ – Minerals and their locations database and information
minerals.net/ – Mineral and gemstone database and information

Websites for guidelines and regulations regarding gems

Convention on International Trade in Endangered Species of Wild Fauna and Flora **cites.org/**
The Kimberley Process Certification Scheme **kimberleyprocess.com/**
Provenance Proof initiative for gemstone traceability **provenanceproof.com/**
The World Jewellery Confederation **cibjo.org/**

Books

Evans, J., 2020. *The History of Synthetic Ruby.* Lustre Gemmology Ltd, UK.
Gübelin, E.J. and Koivula, J.I., 2004, 2005. *Photoatlas of Inclusions in Gemstones (Volume 1 to 3).* Gemological Institute
    of America, California, USA.
Hall, C., 2021. *Gemstones.* Dorling Kindersley Handbooks, London, UK.
Henn, U. and Milisenda, C.C., 2004. *Gemmological Tables.* German Gemmological Association, Idar-Oberstein, Germany.
Hodgkinson, A., 2015. *Gem Testing Techniques.* Valerie Hodgkinson, Scotland, UK.
Hughes, R.W., 2014. *Ruby & Sapphire, A Collector's Guide.* RWH Publishing/Lotus Publishing, Thailand.
Hughes, R.W., Manorotkul, W. and Hughes, E.B., 2017. *Ruby & Sapphire, A Gemologist's Guide.* RWH Publishing/Lotus
    Publishing, Thailand.
O'Donoghue, M., 2005. *Artificial Gemstones.* NAG Press, London, UK.
O'Donoghue, M., 2008. *Synthetic, Imitation and Treated Gemstones.* NAG Press, London, UK.
O'Donoghue, M. (editor), 2008. *Gems: Their Sources, Descriptions and Identification,* 6th edn. NAG Press, London, UK.
Read, P., 1988. *Dictionary of Gemmology,* 2nd edn. Butterworths, London, UK.
Read, P., 2008. *Gemmology,* 3rd edn. NAG Press, London, UK.
Schumann, W., 2006. *Gemstones of the World,* 3rd edn. Sterling Publishing, New York, USA.
Winters, C., 2008. *A Students' Guide to Spectroscopy.* OPL Press, Surrey, UK.

Journals and Gemmological Associations

Accredited Gemologists Association
American Gem Trade Association
The Australian Gemmologist, Gemmological Association of Australia
Canadian Institute of Gemmology
ExtraLapis English/Mineral Monograph series, Lithographie Ltd.
Gems and Gemology, Gemological Institute of America
The Jeweller, The National Association of Jewellers
Journal of Gemmology, The Gemmological Association of Great Britain.
InColor Magazine, International Colored Gemstone Association
International Gem Society

Index

Picture credits

p.4 ©The Steinmetz Diamond Group; p.7 (top), p.88(bottom), p.94, p.96, p.97, p.98 (top right, left) ©Aurora Gems/Photo: Robert Weldon; p.7 (bottom) ©Marco Ansaloni/Science Photo Library,; p.9 ©CRStocker/Shutterstock; p.10 ©USGS; p.11, p.33 (top), p.191 (bottom) ©Fine Minerals International Collection/Photo:Jeff Scovil; p.15 ©hecke61/Shutterstock; p.18, p.95 (top) (bottom)©GIA; p.21 (top right) ©Science Photo Library; p.22 ©Designua/Shutterstock; p.24 (bottom), 26 (bottom left), p.40 (middle), p.58 (right), p.59 (left), p.60, p.61 (top), p.74 (bottom), p.75 (bottom), p.79 (top), p.99, p.109, p.124 (bottom), p.155 (bottom), p.157 (left) Photomicrograph by Nathan Renfro.©GIA; p.25 (bottom), p.74 (top), p.80 (bottom) ©Photo by Robert Weldon; p.34 ©Paul.B.Moore/Shutterstock; p.37 (top) Photomicrograph by Jonathan Muyal and Nicole Ahline ©GIA; p.42 ©Sara Abey, CC BY-SA 4.0, via Wikimedia Commons; p.46 ©Orwin Products Ltd, and as an extract from our Book 'A Student's Guide To Spectroscopy 2nd Edition' by Colin H Winter & Hilary Winter; p.54, p.55(bottom), p.65(top) ©G & R Meieran Collection/Photo:Jeff Scovil; p.55 (top), p.82 (bottom), p.115 (top), p.123 (bottom), p.128 (top) ©Jeff Scovil; p.63 ©Jeremy C Smith Collection/Photo:Jeff Scovil; p.66, p.126 (top) ©Metzger/Cornell/Photo:Jeff Scovil; p.77 (bottom left), p.78 (bottom) ©RW Wise Collection/Photo:Jeff Scovil; p.79 (bottom) ©Dorling Kindersley/UIG/ Science Photo Library; p.81 (bottom), p.84 ©RT Boyd, Ltd Collection/Photo:Jeff Scovil; p.82 (top)p.158 (left) ©Photo by: Ozzie Campos Junior, gemstone cut by John Dyer Gems; p.86 Photo by Shane McClure/©GIA. Courtesy: Grünes Gewölbe, Staatliche Kunstsammlungen Dresden; p.90 (top left) ©Green Mountain Minerals/Photo:Jeff Scovil; p.91 ©Francesco Zerilli/Zerillimedia/Science Photo Library; p.92 (bottom) Photo by Kevin Schumacher ©GIA; p.93 ©Krushevskaya/ Shutterstock; p.95, (bottom left, middle) Aurora Gems; p.95 (bottom,right) ©Dane A. Penland; p.100 (top) ©Photo by Johnny Leung. ©GIA; p.102 (bottom) ©DiamondGalaxy/Shutterstock; p.103 ©P & J Clifford Collection/Photo:Jeff Scovil; p.111 (bottom) ©Fred Parker Collection/Photo:Jeff Scovil; p.119 (top left) ©Lina Jakaite, creative commons https:// linajakaite.gumroad.com/?sort=most_reviewed; p.119 (top right) ©J & M Houran Collection/Photo:Jeff Scovil; p.119 (bottom right) © R & S Jackson Collection/Photo:Jeff Scovil; p.130 (left) ©Ashmolean Museum, University of Oxford/ Heritage Images/Science Photo Library; p.142 © Mineralogy Division, Geological and Planetary Sciences, Caltech (n.d.) CC BY-NC; p.149 (left) ©Pala International Collection/Photo:Jeff Scovil; p.151 (bottom) ©Albert Russ/Shutterstock, p.152 (top) ©Linda Blumel Collection/Photo:Jeff Scovil; p.163 (bottom) ©Ekkehard Schneider/Jeff Scovil, p.166 (top) ©Nika Lerman/ Shutterstock; p.167 ©Jim & Gail Spann Collection/Photo:Jeff Scovil; p.179 ©Mike Keim Collection/Photo:Jeff Scovil; p.183 (bottom) ©Jon Bodsworth; p.191 (top left) ©Ekkehard Schneider Collection/Photo:Jeff Scovil, (top right) ©Barb Dutrow Collection/Photo:Jeff Scovil; p.202 ©Collector's Edge/Photo:Jeff Scovil;p.204 (top) ©Terry Huizing Collection/Photo:Jeff Scovil; p.207 (bottom)© Steve Smale Collection/Photo:Jeff Scovil; p.211 (top) ©H & M Obodda Collection/Photo:Jeff Scovil; p.211 (bottom) ©Rick Kennedy Collection/Photo:Jeff Scovil.

Unless otherwise stated images copyright of Natural History Museum, London.

Every effort has been made to contact and accurately credit all copyright holders. If we have been unsuccessful, we apologise and welcome correction for future editions and reprints.

Author biography

Growing up in her hometown of Perth, Western Australia, Robin developed a love of minerals and geology, completing a BSc with Honours in Geology. She worked for three years in the mining sector before moving to London, UK and settling into a position in the private mineral collector business, working with incredible mineral specimens. During the following decade she complemented this role with a Diploma of Gemmology through the Gemmological Association of Great Britain and was awarded the prestigious Tully Medal. She was thrilled to be offered her dream job as Curator, Minerals and Gemstones at the Natural History Museum, London in 2015, and feels privileged to help care for and manage this magnificent collection of approximately 185,000 specimens.

Acknowledgements

I am indebted to Kerry Gregory (of *Gemmology Rocks*) for her critical review of the manuscript and her many helpful comments. I am grateful to Dr A J A (Bram) Janse for his helpful review of diamonds, Natasha Almeida for her advice regarding tektites and Libyan desert glass, to Andrew Ross and Edmund Jarzembowski for assistance with amber, Alan Hodgkinson for helping with my questions on dispersion, Jeffrey Post and Gabriela Farfan for providing information on specimens held in the Smithsonian Institution collections, and to other colleagues including Eloïse Gaillou and Mike Rumsey for answering my queries. A special debt of gratitude is owed to Leonie Rennie for her many valuable discussions, comments and suggestions.

I would like to thank Jonathan Jackson, Aimee McArdle, Kevin Webb and especially Lucie Goodayle at the Natural History Museum Photo Unit for beautifully photographing many, many gems, and to Jeff Scovil, Nathan Renfro, Robert Weldon, and John Dyer Gems for the use of their wonderful images. I am also grateful to Alan Bronstein for kindly allowing the use of images of his gems, and for his advice on the colour of diamonds.

Many thanks go to the Natural History Museum Publishing team for their patience and guidance for what has turned into a much larger project than planned. And lastly to my friends and colleagues for keeping me going through the writing of 'The Book'. I hope that I have done you all proud.